T0191828

Biomaterials Science and Implants

Bikramjit Basu

Biomaterials Science and Implants

Status, Challenges and Recommendations

Springer

Bikramjit Basu
Materials Research Center and BioSystems
Science and Engineering
Indian Institute of Science Bangalore
Bengaluru, Karnataka, India

ISBN 978-981-15-6920-3 ISBN 978-981-15-6918-0 (eBook)
https://doi.org/10.1007/978-981-15-6918-0

This Springer imprint is published by the registered company Springer Nature Singapore Pte Ltd.
The registered company address is: 152 Beach Road, #21-01/04 Gateway East, Singapore 189721, Singapore

Foreword by The President, Indian National Science Academy

The Indian National Science Academy has assigned itself an important mandate of informing society, stakeholders and policymakers on various issues of science that are of relevance to society. This is being carried out through comprehensive reports and books, on the status of specific subjects through a narrative of evidence-based understanding of scientific research at the cutting edge. The literature is presented in a simple to understand manner and deals with aspects of education, research, and future possibilities, future challenges and suggestions for policies to take India forward. These reports project the needs of India during the coming two to three decades.

This monograph on *Biomaterials Science and Implants* by Prof. Bikramjit Basu presents one such effort. The book deals with a complimentary subject of biomaterials, with the present science and education scenario, traces the historical developments, summarises the present status and gazes in to the future. It discusses various realms of this science that he terms as immigration science, that derives/depends on contributions from a variety of disciplines, such as material sciences, especially the synthetic hybrid materials, biological sciences, health sciences, toxicological research besides biomechanics and manufacturing sciences. Various chapters of the monograph deal with (a) important developments, especially in the field of medical implants, the need to augment teaching in this discipline and the future possibilities, given that biomaterials will be an important economic driver for the country; (b) current research scenarios and technical challenges; and (c) the need for future research. The monograph is well-illustrated with data and figures to help the reader. This monograph was reviewed by two experts and we thank them for their help in providing timely and incisive reviews. Both the experts commended this monograph.

On behalf of INSA, I am pleased to present this monograph to the public and policymakers, including the scientific fraternity, and I am sure it will eminently serve the purpose, it was meant for. I would like to place on record our appreciations for Prof. M. S. Valiathan to write a Foreword for this monograph. His doing so, speaks volumes on the quality of science being presented by the author. I compliment and congratulate Prof. Basu for his hard work. I sincerely hope that this book will enthuse newer communities to join hands to develop major research initiatives in this important area of biomaterials and bioengineering and create nationwide traction.

I thank Prof. A. K. Singhvi, Vice President, INSA, for his active role in bringing this book to fruition. I also would like to place on record my appreciation for Prof. Gadadhar Misra for coordinating the publication of this book with M/s Springer Nature India. This is the first book being published and coproduced by M/s Springer Nature.

Ajay K. Sood
FRS, President
Indian National Science Academy
New Delhi, India

Foreword by M. S. Valiathan

In the last couple of decades, the field of biomaterials science and tissue engineering is at the frontier of research and innovation, considering the number of scientific discoveries and their potential impact in treating human diseases. Overall, this book emphasises the enormous need for the supply of regenerated organs and tissues, as the spontaneous capacity for regeneration is limited in the human body. To substantiate the relevance for human healthcare, degenerative and inflammatory problems of bone and joints affect millions of people worldwide.

In order to address biomedically relevant challenges in orthopaedics as well as neural and cardiovascular diseases, researchers must blend the fundamental concepts of engineering sciences (materials science), basic sciences (chemistry and physics) and biological sciences (cell and molecular biology) to engineer synthetic tissue replacements and develop novel healing strategies. Such an interdisciplinary research approach requires understanding across the boundary of remotely linked scientific disciplines. Researchers can develop innovative ideas, as well as understand the language of this important research area of societal relevance. In most significant terms, this monograph closes with the author's recommendations for the policymakers, science administrators and lawmakers to facilitate the growth of this field in near future and long term.

I am convinced that this important monograph, of considerable relevance to India, should inspire many active researchers as well as young researchers, to think laterally, while blending sufficient knowledge of biological systems with engineering sciences to develop biomedical materials. The accelerated growth in the fields of biomaterials and biomedical engineering, when most, if not all, the recommendations are translated to reality, is expected to facilitate affordable, yet high-quality healthcare for millions of Indians and citizens of other countries globally. To accompany the advances, the right regulatory and legislatory changes are required. I personally find this much-needed monograph timely and immensely valuable for the interdisciplinary scientific community pursuing the field of biomaterials and biomedical engineering.

Manipal, India
October 2019

M. S. Valiathan
Ch.M., FRCS, FRCS(C), FRCP, D.Sc (h.c)
National Research Professor

Padma Vibhushan Awardee (2005)

Former Vice-Chancellor
Manipal University

Former President, Indian National Science Academy

Former Director
Sree Chitra Tirunal Institute for Medical Sciences and Technology
Thiruvananthapuram

Preface

Who, What, When, Where, Why and How

We are now in an era of interdisciplinary research, as a matter of necessity and also of knowledge: in themes and topics, in perspectives and problem-definition, in execution and evaluation. The field of biomaterials science is one such interdisciplinary research field. In this monograph, I have made extensive efforts to sketch the Indian landscape at work in this socially important area of scientific research, against the international backdrop. The main aim of this monograph is to stimulate actions by a wide variety of stakeholders who read this report.

I am a materials scientist by academic training and a biomaterials scientist by calling. At the initial phase of my career in biomaterials, I was aware of the challenges involved in entering an unexplored interdisciplinary domain that promises to have a dramatic impact on critical real-world problems in the next decades. I can still recollect the uphill battle when I started pursuing science-to-implant innovation, involving clinicians and companies. Against this backdrop, I have summarised the major challenges and opportunities in the field, for young as well as established researchers in the field.

The Indian medical device market is dominated by foreign companies, with over 70% of demands being met through imports. Currently, approximately one million patients need prostheses and implants every year, and India imports a big slice, worth Rs. 7000 crore a year (approximately $1 billion USD). Alongside, biomaterials research internationally is progressing in leaps and bounds at several institutions of repute. In this context, I have penned down a set of recommendations, which I believe should be considered as priority recommendations by all stakeholders for building the research ecosystem of India's tomorrow.

The Roadmap: Engaging Key Stakeholders

When we define the progress of a particular field, we are required to think of its issues, challenges and bottlenecks. However, equally important is attempting to solve these problems, brainstorm recommendations and action plans.

It is high time that a monograph on the national status of research in biomaterials is presented. It is time for real inquiry into select global institutes of importance, for highlighting challenges and, most of all, issuing actionable recommendations that will expedite inventions from the bench to the bedside. There is no one road-map that fits all ecosystems, and therefore, this monograph is best used as a means to stimulate thinking in the right direction. The process of building dynamic innovation ecosystems involves not only working out the details, but also mastering the art of motivation, organisation and negotiation. It is hoped that this text will be well-received by science administrators, policymakers, legislators and citizens, in the context of developing countries, taking India as an example.

The monograph, however, also contains valuable information for anyone who is interested in learning about the current issues and trends, gaps and synergies, in biomaterials sciences and bioengineering. It should also be useful for those interested in knowing about the status of major national and international research centres. The list of global institutes of importance presented here is not exhaustive and I acknowledge that there are others that have not been highlighted here that are also of high calibre.

In particular, this monograph is visualised as a strategic roadmap to balance the needs of the following stakeholders, towards building the regulatory framework:

1. Science administrators
2. Policymakers and legislators
3. Research coordinators, incubators and funding officers
4. Scientists and young researchers, including Ph.D. students, in Academia
5. MSMEs and industry
6. Clinicians

The Guide: Navigating the Monograph

Readers are encouraged to navigate this report by first considering the Contents, what the purport of each chapter is and then, identifying the sections that are of greatest importance to the reader, based on their expertise, interests and organisation. The first four chapters are intended to provide a foundational discussion of the field and the status of ongoing research, nationally and globally. Subsequent to this, the fifth chapter, “A Challenging Frontier”, is intended to provide the reader with an overview of the key challenges in the field. The sixth chapter, “Recommendations”, is intended to stimulate the minds of the readers on key recommendations and the accompanying timeline-specific action plans.

Actionable Takeaways: Determining What Actions to Take

The main aim of this monograph is to stimulate actions by a wide variety of stakeholders who read this report. The actions should be focussed on contributing to the development of the biomaterials and bioengineering fields. This can be in the form of actions related to specific challenges and recommendations presented here. The list, however, is not exhaustive, and the reader is welcome to think of further challenges and recommendations that can be addressed. I am hopeful that the monograph will trigger new conversations and discussions among peer groups towards a more detailed action plan. It is strongly recommended that the policy-makers and lawmakers make note of actions based on the selected sections of relevance, in the fifth and sixth chapters. I am confident that the insights gained from this monograph, and the actions that it will stimulate, will enable a more robust and congruent ecosystem for biomaterials science and implants. All the views expressed in this monograph by the author are intended solely for the purpose of discussion, and are not intended to offend any sentiments of any individual, research group or other stakeholders in the biomaterial implants ecosystem. The readers are welcome to share their comments or feedback with the author.

A Note of Thanks

Many challenges and recommendations, summarised in this book, are the reflection of my understanding gained during the significant interactions with numerous collaborators from academia, national laboratories, medical institutions and industry as well as young researchers, in India and abroad. I have acknowledged this in a separate list, which also contains a number of science administrators or policy-makers. I am grateful to all of them for their time and useful suggestions. I thank my current and former students, including Nitu Bhaskar, Srimanta Barui, Subhadip Basu, Swati Sharma, Vidushi Sharma, Ranjith Kumar P., Asish Kumar Panda, Gowtham N. H., Soumitra Das, Deepa Mishra, Sulob Roy Chowdhury, Indu Bajpai, Subhadip Bodhak, Yashoda Chandorkar, Subhomoy Chatterjee, Ashutosh K. Dubey, Shilpee Jain, Ravikumar K., Alok Kumar, Prafulla K. Mallik, Atiar Molla, Shekhar Nath, Shibayan Roy, Naresh Saha, Debasish Sarkar, B. Sunilkumar, Greeshma T., Garima Tripathi and Raghunandan Ummethala. I deeply acknowledge the support of Dr. Nandita Keshavan, Dr. Damayanti Datta, Mrs. Prerana S., Ms. Sheetal Chowdhury, Ms. Titash Mukherjee, Ms. Rea Johl and Mr. Nihal Kottan in preparing this monograph. I also appreciate the comments and constructive suggestions of the reviewers of this book. Finally, I am grateful to my wife, Pritha Basu, and son, Prithvijit Basu, as well as other family members for their unconditional support during writing this monograph. I am also indebted to my parents, Manoj Mohan Basu (father) and Chitra Basu (mother) for their constant encouragement and inspiration throughout my life. I truly acknowledge the financial support received from the Indian National Science Academy, Scheme for Promotion of Academic and Research Collaboration (SPARC) and Indian Institute of Science, Bangalore, during the writing of this monograph.

Bikramjit Basu
Ph.D., CEng., FACerS, FAMS, FAIMBE, FNAE, FNASc, FAScT, FBAO
Professor
Materials Research Center and BioSystems
Science and Engineering
Indian Institute of Science, Bangalore
Bengaluru, Karnataka, India

Honorary Professor
University of Manchester
Manchester, UK

Acknowledgements

List of Scientists from Academia and National Laboratories Consulted

Kantesh Balani, Indian Institute of Technology Kanpur, India
Rinti Banerjee, Indian Institute of Technology Bombay, India
Dipankar Banerjee, Indian Institute of Science, Bangalore, India
Ananya Barui, Indian Institute of Engineering Science and Technology, Shibpur, India
Naresh Bhatnagar, Indian Institute of Technology Delhi, India
Aldo Boccaccini, Institute of Biomaterials, Erlangen, Germany
K. Muraleedharan, CSIR-CGCRI, Kolkata, India
Vamsi Krishna Balla, CSIR-CGCRI, Kolkata, India
Subhadip Bodhak, CSIR-CGCRI, Kolkata, India
Biswanath Kundu, CSIR-CGCRI, Kolkata, India
Marc Bohner, RMS Foundation, Bettlach, Switzerland
Rajendra K. Bordia, Clemson University, USA
Annabel Braem, Katholieke Universiteit Leuven, Belgium
Sarah Cartmell, University of Manchester, UK
Kaushik Chatterjee, Indian Institute of Science, Bangalore, India
Jérôme Chevalier, Institut National des Sciences Appliquées (INSA), Lyon, France
Pallab Dutta, Indian Institute of Engineering Science and Technology, Shibpur, India
Mitun Das, CSIR-CGCRI, Kolkata, India
Brian Derby, University of Manchester, UK
Alok Dhawan, CSIR-Indian Institute of Toxicology Research, Lucknow, India
Santanu Dhara, Indian Institute of Technology Kharagpur, India
Christophe Drouet, Centre Inter-universitaire de Recherche et d'Ingénierie des Matériaux, Toulouse, France
Ashutosh K. Dubey, Indian Institute of Technology (BHU), Varanasi, India

Michael Gelinsky, Technische Universität Dresden, Germany
Liesbet Geris, Katholieke Universiteit Leuven, Belgium
Sourabh Ghosh, Indian Institute of Technology Delhi, India
Swati Haldar, Indian Institute of Technology Roorkee, India
Julian Jones, Imperial College London, UK
Surya Kalidindi, Georgia Institute of Technology, USA
Subramani Kanagaraj, Indian Institute of Technology Guwahati, India
Manoj Komath, Sree Chitra Tirunal Institute for Medical Sciences and Technology, Thiruvananthapuram, India
Veena Koul, Indian Institute of Technology Delhi, India
Ashok Kumar, Indian Institute of Technology Kanpur, India
A. M. Kuthe, Visvesvaraya National Institute of Technology, Nagpur, India
Debrupa Lahiri, Indian Institute of Technology Roorkee, India
Cato Laurencin, University of Connecticut, USA
Anne Leriche, Université Polytechnique Hauts-de-France, Valenciennes, France
H. S. Maiti, Former Director, CSIR-Central Glass and Ceramic Research Institute, Kolkata, India
Biman B. Mandal, Indian Institute of Technology Guwahati, India
Saumen Mandal, National Institute of Technology Karnataka, Surathkal, India
Sujata Mohanty, All India Institute of Medical Sciences, New Delhi, India
Subha Narayan Rath, Indian Institute of Technology Hyderabad, India
Abhay Pandit, National University of Ireland, Galway, Ireland
Hardik Pandya, Indian Institute of Science, Bangalore, India
Jouni Partanen, Aalto University, Finland
Falguni Pati, Indian Institute of Technology Hyderabad, India
Yarlagadda Prasad, Queensland University of Technology, Australia
Seeram Ramakrishna, National University of Singapore, Singapore
B. Ravi, Indian Institute of Technology Bombay, India
Rui Reis, 3Bs' Research Group, University of Minho, Braga, Portugal
Amit Roy Chowdhury, Indian Institute of Engineering Science and Technology, Shibpur, India
Debashish Sarkar, National Institute of Technology, Rourkela, India
T. S. Sampathkumar, Indian Institute of Technology Madras, India
Jukka Seppälä, Aalto University, Finland
C. P. Sharma, Sree Chitra Tirunal Institute for Medical Sciences and Technology, Thiruvananthapuram, India
D. D. Sarma, Indian Institute of Science, Bangalore, India
Molly Shoichet, University of Toronto, Canada
Carl G. Simon, Jr., National Institute of Standards & Technology (NIST), Gaithersberg, MD, USA
Neetu Singh, Indian Institute of Technology Delhi, India
Simone Sprio, Istituto di Scienza e Tecnologia dei Materiali Ceramici (ISTEC), Faenza, Italy

N. Ravi Sundaresan, Indian Institute of Science, Bangalore, India
Anna Tampieri, Istituto di Scienza e Tecnologia dei Materiali Ceramici (ISTEC), Faenza, Italy
H. K. Varma, Sree Chitra Tirunal Institute for Medical Sciences and Technology, Thiruvananthapuram, India

List of Clinicians Consulted

Aniruddh T. J., M. S. Ramaiah Memorial Hospital, Bangalore, India
Yogesh Chawla, Postgraduate Institute of Medical Education and Research, Chandigarh, India
Amit. K. Dinda, All India Institute of Medical Sciences, New Delhi, India
Anil Mandhani, Medanta, the Medicity, Gurugram, India
Ivan Martin, University Hospital of Basel, Switzerland
K. V. Menon, Sparsh Hospital, Bangalore, India
B. V. S. Murthy, Ramaiah University of Applied Sciences, Bangalore, India
Sunil Nikose, Datta Meghe Institute of Medical Sciences, Wardha, India
Zahir Quaziuddin, Datta Meghe Institute of Medical Sciences, Wardha, India
C. Rex, Rex Hospital, Coimbatore, India
Vibha Shetty, Ramaiah University of Applied Sciences, Bangalore, India
Balendra Singh, King George's Medical University, Lucknow, India
D. C. Sundaresh, Ramaiah University of Applied Sciences, Bangalore, India
Rajesh T. R., Sparsh Hospital, Bangalore, India
M. S. Valiathan, Manipal University, India
Ajit Yadav, Gleneagles Global Health City, Chennai, India

List of Science Administrators/Policymakers Consulted

Florent Bernard, European Commission, Brussels, Belgium
Kakali Dey Dasgupta, Department of Biotechnology, Government of India, New Delhi, India
Katherine Freeman, Healthcare Technologies, Engineering and Physical Sciences Research Council (EPSRC), UK
Kalaivani Ganesan, Department of Biotechnology, Government of India, New Delhi, India
Priyankana Mukherjee, IKP Engineering, Design, and Entrepreneurship Network, Bangalore, India
Gert Roebben, European Commission, Brussels, Belgium

Alka Sharma, Department of Biotechnology, Government of India, New Delhi, India
Ashutosh Sharma, Department of Science and Technology, Government of India, New Delhi, India
Jitendar Sharma, Andhra Pradesh MedTech Zone, India
Soumya Swaminathan, World Health Organisation
Sandeep Verma, Science Engineering and Research Board, Government of India, New Delhi

List of Industry Professionals Consulted

Rohan Aggarwal, Vidcare Innovations, Pune, India
S. K. Banerji, Orthotech, Gujarat, India
Debasish Bhattacharjee, TATA Steel Limited, India
Sudip Bose, TATA Steel Limited, India
M. Chandrashekharan, SMATEC b.v.b.a., Belgium
Aroop K. Dutta, Excel Matrix Biological Devices Pvt. Ltd., Hyderabad, India
Ranjna C. Dutta, Excel Matrix Biological Devices Pvt. Ltd., Hyderabad, India
Nilay Lakhkar, SynThera Biomedical, Pune, India
Subrata Mukherjee, TATA Steel Limited, India
Kamlesh Patel, Kaaryans Technical Ceramics, Gujarat, India
Kingshuk Poddar, TATA Steel Limited, India
Sabyasachi Roy, ANTS Ceramics, Vasai, India
Ravi Sarangapani, Biomedical Engineering Consultant, Pune, India
Mario van Wingerde, TATA Steel Limited, India

List of Institutions Consulted

Andhra Pradesh Med Tech Zone, Visakhapatnam, India
BETiC (Biomedical Engineering and Technology Incubation Centre), India
IKP Engineering, Design, and Entrepreneurship Network, Bangalore, India
Istituto di Scienza e Tecnologia dei Materiali Ceramici (ISTEC), Faenza, Italy
Kalam Institute of Health Technology, Visakhapatnam, India
PSG Institute of Advanced Studies, Coimbatore, India
Sree Chitra Tirunal Institute for Medical Sciences and Technology, Thiruvananthapuram, India

List of Funding Agencies

Department of Biotechnology, Government of India
Department of Science and Technology, Government of India
Indian National Academy of Engineering, Gurugram, India
Indian National Science Academy, New Delhi, India
Scheme for Promotion of Academic and Research Collaboration, Government of India

Executive Summary

"Nothing is more powerful than an idea whose time has come".

Do you know what hides in your hospital bills? Take a deep breath and prepare to be surprised: 40% of the expenses are attributed to medical equipment, of which about 70% come from high-tech diagnostic tests.

Medical devices single-handedly offset the healthcare cost impact. That is because, nearly 80% of devices come from outside the country: the USA, Germany, France, Singapore, China and the Netherlands. The dark side of the story is well-known: one in every four households in India is pushed into debt every year over healthcare costs. Not just Indians, consider the large number of patients who come for treatment from the Middle East, Africa and neighbouring countries (Bangladesh, Sri Lanka, Nepal, Bhutan, etc.).

The deficits of the current research ecosystem in India play a key role in this chain of skyrocketing costs and unmet clinical needs. A key challenge for innovation is the manufacturing of affordable biomedical devices without compromising on quality. Made of metals and alloys, ceramics and carbons, polymers and composites and other materials, biomaterials are used every day in surgery, in dental applications and orthopaedics and drug delivery. Then, there are smart biomaterials, a field that is developing at a very rapid pace, which can interact with biological systems, directly influencing cell behaviour.

India has sent out tectonic ripples across the world with its mammoth healthcare scheme for the under-privileged citizens. In line with the vision of affordable medical costs, it is essential to develop a national roadmap for indigenous high-performance biomedical implants and devices, in an accelerated manner. To bring the benefits of cutting-edge science from the laboratory benchside to the patient bedside, it is imperative to complete the full cycle of translational research. For all persons involved in this field, from all sectors, it is the right time to think about implementing engaging action plans that are tailored to the needs of the times. In order to introduce the readers to the status of the field, the introductory chapter discusses important concepts in the field, and the Appendix lists internationally approved definitions as well as details of the national meetings, which led the author to conceive this text.

A major bottleneck is the poorly conceived regulatory framework in India. The strategic road ahead lies in implementing robust national regulatory policy and translational research programmes with strong involvement of clinicians and industry, towards the discovery and deployment of new biomaterials and implants. Furthermore, the author proposes the development of biomaterialomics (biomaterials with data science), which relies on a data-driven integrated understanding of biocompatibility and elements of biomaterials development, while leveraging both conventional and advanced manufacturing (e.g. 3D bioprinting).

The recommendations laid down in this monograph should help to establish India as a global market leader for a new generation of bioimplants, whose predictive clinical performance would be closely tracked by "digital twins". An outcome of this strategic initiative would be realised in the creation of new employment opportunities for skilled and semi-skilled Indians, together with the education and training of next-generation researchers.

It is the author's vision that such a text would not only encourage young researchers to be passionate about understanding the current challenges, and those of the end-user, but also inspiring them to form strong collaborations for adaptive problem solving, thereby making significant contributions to the field. Also, this monograph would be useful in honing the knowledge of those established in the field by broadening their understanding on the research ecosystem in developed nations and the national context. These topics are of greater importance towards building an India of the future, where innovations can have a palpable impact, and technological innovations meet the healthcare needs of society. Given the right support, the key recommendations presented here can be highly transformative to the field. "Nothing can stop an idea whose time has come", Victor Hugo, the French epic novelist, once wrote. We have proposed here the idea of the moment. An idea that can carry forward both people's wellbeing and the nation's economy in its energy and momentum.

Contents

About the Author

http://bikramjitbasu.in/

Professor Bikramjit Basu is currently a Professor at the Materials Research Center and holds Associate Faculty position at Center for Biosystems Science and Engineering, Indian Institute of Science (IISc), Bangalore. He is currently Visiting Professor at University of Manchester, UK. After his undergraduate and postgraduate degree in Metallurgical Engineering from NIT Durgapur and IISc respectively, he earned his PhD in the area of Engineering Ceramics at Katholieke Universiteit Leuven, Belgium in March, 2001. Following a brief post-doctoral stint at University of California, Santa Barbara; he served as a faculty of Indian Institute of Technology Kanpur during 2001-2011. He has taught in UK, Spain, Slovenia, Belgium and Nepal. He has successfully led international research programs with USA, UK and Germany.

Professor Basu has been pursuing challenging interdisciplinary research at the cross-road, where Materials Science, Biological Science and Medicine meet. He has aptly used the principles of Biomaterials Science and Biomedical Engineering to develop next generation implants and biomedical engineering solutions in an effort to address unmet clinical needs for musculoskeletal, dental, neurosurgical and urological applications. Over the years, he has created interactive and intensive collaborations with a number of clinicians and entrepreneurs to accelerate bench science-to-device prototype development. Encompassing theoretical predictions, computational analysis, experimental discovery and clinical translational research, his research group has laid the foundation for intelligent design of implants, 3D binderjet printing of biomaterials, science of biocompatibility and bioengineering strategies, to advance the development of biomedical implants, regenerative engineering and bioelectronic medicine. He is currently leading India's major translational Center of Excellence on biomaterials with an interdisciplinary team of researchers from academia and medical institutions and hospitals. He has published over 300 peer-reviewed research papers in leading journals (total citations ~ 11,000 and H-index: 56). Many of his former students are currently thriving research programs in IITs and NITs in India. He has co-authored 9 books, including 7 textbooks on Biomaterials, Tribology and Ceramics.

Prof. Basu's contributions in Biomaterials Science have been widely recognised. He received India's most coveted science and technology award, Shanti Swarup Bhatnagar Prize in 2013. A Chartered Engineer of UK, he is an elected Fellow of the International Academy of Medical and Biological Engineering (2020), International Union of Societies for Biomaterials Science and Engineering (2020), Indian Academy of Sciences (2020), American Ceramic Society (2019), American Institute of Medical and Biological Engineering (2017), Institute of Materials, Minerals & Mining, UK (2017), National Academy of Medical Sciences, India (2017), Indian National Academy of Engineering (2015), Society for Biomaterials and Artificial Organs (2014) and National Academy of Sciences, India (2013). He is currently serving as Advisor to TATA Steel New Materials Business and is an Abdul Kalam National Innovation Fellow.

Abbreviations

3D	Three-dimensional
3DP	Three-dimensional printing
3DPL	Three-dimensional plotting
3DPP	Three-dimensional powder printing
AB-PMJAY	Ayushman Bharat Pradhan Mantri Jan Arogya Yojana
AcE	Accelerating Entrepreneurs
ADMI	Association of Diagnostics Manufacturers of India
AERB	Atomic Energy Regulatory Board
AI	Artificial Intelligence
AIG	Asian Institute of Gastroenterology
AIIMS	All India Institute of Medical Sciences
AiMeD	Association of Indian Medical Devices Industry
AIR	Academic Innovation Research
AJRR	American Joint Replacement Registry
AMCHAM	American Chamber of Commerce in India
AMTZ	Andhra Pradesh Med Tech Zone
APIs	Application Programme Interfaces
ARC	Australian Research Council
BARC	Bhabha Atomic Research Centre
BCIL	Biotechnology Consortium of India Limited
BCP	Biphasic Calcium Phosphate
BETiC	Biomedical Engineering and Technology Incubation Centre
BHU	Banaras Hindu University
BIPP	Biotechnology Industry Partnership Programme
BIRAC	Biotechnology Industry Research Assistance Council
BIS	Bureau of Indian Standards
BME	Biomedical Engineering
BRICS	Brazil, Russia, India, China and South Africa (Five major emerging economies)
BRIT	Board of Radiation and Isotope Technology

CAGR	Compound Annual Growth Rate
CASPA	Calcium-Sulfate-Phosphate Active Composition
CBE	Cellular and Biochemical Engineering
CDSCO	Central Drugs Standard Control Organisation
CE mark	Certification mark for products within the European Economic Area
CGCRI	Central Glass and Ceramic Research Institute
CII	Confederation of Indian Industry
CIPET	Central Institute of Plastics Engineering & Technology
CLRI	Central Leather Research Institute
CMC	Christian Medical College
CMF	Craniomaxillofacial
CRO	Contract Research Organisation
CRS	Contract Research Scheme
CSIR	Council of Scientific and Industrial Research
CT	Computerised tomography
DAE	Department of Atomic Energy
DBT	Department of Biotechnology
DCC	Drugs Consultative Committee
DEITY	Department of Electronics and Information Technology
DRDO	Defence Research and Development Organisation
DST	Department of Science and Technology
DTAB	Drugs Technical Advisory Board
ECM	Extracellular matrix
EDEN	Engineering, Design and Entrepreneurship Network
EIR	Entrepreneur-in-residence
EMC	Electro-Magnetic Compatibility testing
EMI	Electro-Magnetic Interference testing
ENT	Ear Nose Throat
EPSRC	Engineering and Physical Sciences Research Council
EXAFS	X-ray Absorption Spectroscopy
FDA	Food and Drug Administration
FICCI	Federation of Indian Chambers and Commerce Industry
GATE	Graduate Aptitude Test in Engineering
GCE	Grand Challenges Exploration
GLP	Good Laboratory Practice
GMP	Good Manufacturing Practice
GYTI	Gandhian Young Technological Innovation
HA	Hydroxyapatite
HDPE	High-Density Polyethylene
HTA	Health Technology Assessment
IBSC	Indian Biomedical Skill Certificate
ICME	Integrated Computational Materials Engineering
ICMR	Indian Council of Medical Research
ICSSR	Indian Council for Social Science Research

IIEST	Indian Institute of Engineering Science and Technology
IISc	Indian Institute of Science
IISER	Indian Institute of Science Education and Research
IIT	Indian Institute of Technology
IKP Trust	Innovation, Knowledge, Progress Trust
IMDRF	International Medical Device Regulators Forum
IMPRINT	IMPacting Research Innovation and Technology
IOP	Intraocular Pressure
IP	Intellectual Property
IPR	Intellectual Property Rights
ISO	International Organisation for Standardisation
ISTEC	Istituto di Scienza e Tecnologia dei Materiali Ceramici
IVRI	Indian Veterinary Research Institute
JIPMER	Jawaharlal Institute of Postgraduate Medical Education
KIHT	Kalam Institute of Health Technology
LSRB	Life Science Research Board
MADE in SC	Materials Assembly and Design Excellence in South Carolina
MCI	Medical Council of India
MD	Molecular Dynamics
MDA	Medical Devices Authority
MDTAG	Medical Devices Technical Advisory Group
MEDHA	Medical Device Hackathon
MEDIC	Medical Device Innovation Camp
MHRD	Ministry for Human Resource Development
MIILI	Make In India leadership Institute
ML	Machine Learning
MOU	Memorandum of Understanding
MRI	Magnetic Resonance Imaging
MSCs	Mesenchymal Stem Cells
MSMEs	Micro, Small and Medium scale Enterprises
MSRUAS	M. S. Ramaiah University of Applied Sciences
MTAI	Medical Technology Association of India
MWCNT	Multi-Walled Carbon Nanotubes
NATFOS	National Frontiers of Science
NHS	National Health Stack
NHSRC	National Health Systems Resource Centre
NIB	National Institute of Biologicals
NIH	National Institutes of Health
NIPUN	Non-regulatory Innovation Potential Utility and Novelty Certificate
NIST	National Institute of Standards & Technology
NIT	National Institute of Technology
NML	National Metallurgical Laboratory
NPPA	National Pharmaceutical Pricing Authority
NSF	National Science Foundation

NSTIF	National Science, Technology & Innovation Foundation
PACE	Promoting Academic Research Conversion to Enterprise
PANI	Polyaniline
PCL	Polycaprolactone
PCT	Patent Cooperation Treaty
PDMS	Polydimethylsiloxane
PFMEA	Process Failure Mode Effects Analysis
PGA	Polyglycolic Acid
PGIMER	Postgraduate Institute of Medical Education and Research
PLA	Polylactic Acid
PLGA-CNF	Polylactic-Polyglycolic Acid-Carbon nanofibre
PM-RSSM	Pradhan Mantri-Rashtriya Swasthya Suraksha Mission
PM-STIAC	Prime Minister's Science, Technology and Innovation-Advisory Council
PSP	Process-Structure-Property
PTH	Parathyroid Hormone
PVA	Polyvinyl Alcohol
PVDF-CNT	Polyvinylidene Fluoride-Carbon Nanotube
QA	Quality Assurance
QC	Quality Control
QCI	Quality Council of India
R&D	Research and Development
RCT	Randomised Controlled Trial
ROME	Reorientation of Medical Education
SBIRI	Small Business Innovation Research Initiative
SCTIMST	Sree Chitra Tirunal Institute for Medical Sciences and Technology
SEARN	South East Asia Regulatory Network
SEED Fund	Sustainable Entrepreneurship and Enterprise Development Fund
SERB	Science and Engineering Board
SF	Silk Fibroin
SGPGI	Sanjay Gandhi Postgraduate Institute of Medical Sciences
SIB	School of International Biodesign
SIIHEI	Start-up India Initiative for Higher Education Institutions
SME	Small-to-Medium Enterprise
SPARC	Scheme for Promotion of Academic and Research Collaboration
SPARSH	Social Innovation programme for Products: Affordable and Relevant to Societal Health
SRISTI	Society for Research and Initiatives for Sustainable Technologies and Institutions
SSSIHMS	Sri Sathya Sai Institute of Higher Medical Sciences
STEM	Science, Technology, Engineering and Mathematics
SUPRA	Scientific and Useful Profound Research Advancement
TCP	Tricalcium Phosphate
TCS	TATA Consultancy Services
TERM	Tissue Engineering and Regenerative Medicine

THA	Total Hip Arthroplasty
THR	Total Hip Joint Replacement
THSTI	Translational Health Science and Technology Institute
TIMed	Technology Business Incubator for Medical Devices and Biomaterials
TKR	Total Knee Replacement
TMJ	Temporomandibular Joint
TRL	Technology Readiness Level
UAY	Uchchatar Avishkar Yojna
UHMWPE	Ultra-high Molecular Weight Polyethylene
USA	United States of America
USD	United States of America Dollar
WFIRM	Wake Forest Institute for Regenerative Medicine
WHO	World Health Organisation
ZTA	Zirconia Toughened Alumina

Chapter 1
Crossing the Boundaries

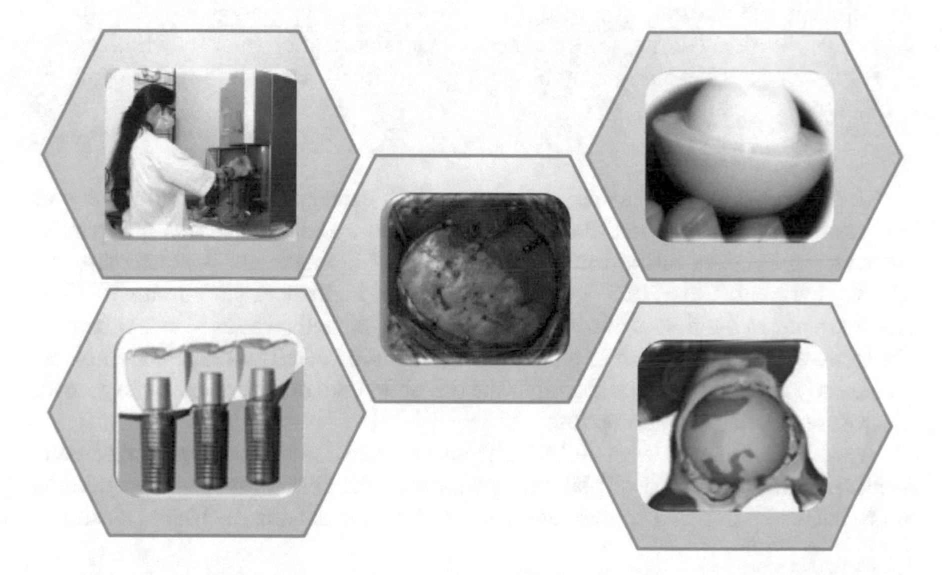

Abstract The future of science lies in breaking the down barriers between fields to solve global challenges of our time. 'Interdisciplinarity' is the mantra to solve the grand global challenges that no single discipline can address on its own. Scientists from various backgrounds are coming together to transcend the boundaries of their respective disciplines. The most exciting research is happening at the intersections, where various scientific disciplines meet. The field of biomaterials science is one of the finest examples of a truly interdisciplinary research field. This chapter introduces the readers to key definitions of the scientific terms that a non-specialist needs to know, before appreciating the major developments that have taken place over the last few decades. The relevance of this important interdisciplinary field of science is illustrated by citing a few biomaterials-based solutions to treat human diseases.

B. Basu, *Biomaterials Science and Implants*,
https://doi.org/10.1007/978-981-15-6918-0_1

1.1 Biomaterials Bloom

In the recent past, there has been significant attention to bridge the gap between biology and engineering sciences. Consistent efforts have resulted in the conceptual evolution of 'bioengineering' or 'biomedical engineering' or 'clinical engineering', as a distinct discipline. Another example is biomaterials science, which bridges the gap between materials science and biology (see Fig. 1.1). The readers can refer to a set of definitions in Appendix A. If disciplines are related, then 'crossing the boundary' becomes a little easier, but it is more challenging, when they are disconnected.

> Although biomaterials as a field, formally, came into existence only after the Second World War, the development of this important class of materials has been well documented throughout history, albeit any scientific assessment of their biocompatibility.

The ancient Indian physician, Susruta of sixth-century BC, reported the use of sutures made of human hair, flax, and hemp to close skin incisions in his treatise, *Suśruta-samhitā*. As far as the early evolution of biomaterials is concerned, the Greek physician, Galen of Pergamum, has also described catgut sutures to close large wounds in the first-century AD. Nacre from sea shells was used as a substitute for teeth, while metals, such as gold and wrought iron, were used as dental implants in the early ages. Such cases highlight the use of an implant or prosthesis to recover the loss of an anatomical function.

As of 1829, Henry Levert of Alabama investigated 21 different types of metallic sutures, with those made of platinum performing the best. Surgeons then became more interested in metal screws and plates to fix bone defects. In 1886, Hansmann

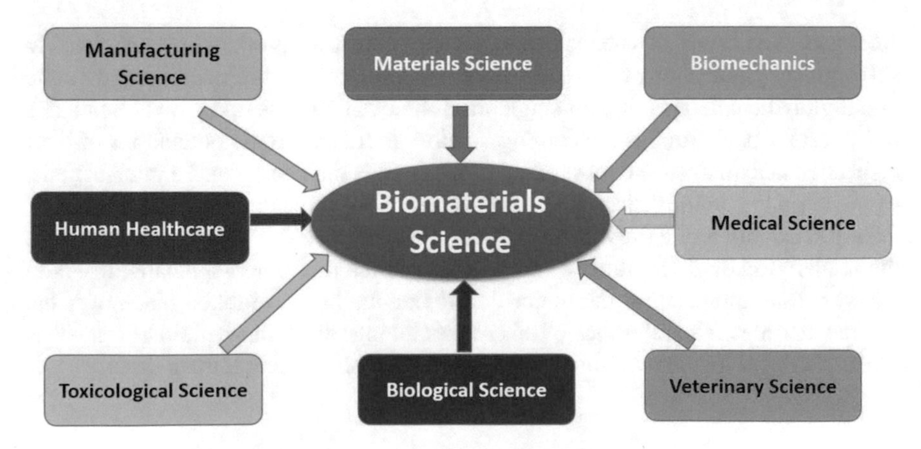

Fig. 1.1 Biomaterials science: integrates concepts from multiple disciplines

used plates and screws successfully for the first time in Germany. In 1881, Etienne-Jules Marcy published the first design for an artificial heart. By 1926, a new formulation of stainless steel 18-8 has been formulated, which helped to address the issue of corrosion of iron and steel in the body. It took around 100 years before anything similar would be implanted into a human patient. The first blood vessel replacement, using parachute material, was reported in 1952. This was found by Arthur Zierold in metallic suture experiments.

The field of biomaterials blossomed mainly after the Second World War due to the large need of the soldiers, who sustained battlefield injuries. Polymethyl methacrylate (PMMA) shards, that were accidentally lodged in the eyes of aviators, did not create inflammation or an adverse tissue reaction. This inspired British ophthalmologist, Sir Harold Ridley, to create early intraocular lenses out of polymethyl methacrylate. Today, polydimethylsiloxane (PDMS) lenses are being used instead, and cataract-associated blindness can be avoided with the use of PDMS lenses.

A fine example of the use of a biomaterial for the treatment of musculoskeletal diseases is that of the total hip joint replacement (THR), which has evolved to become one of the most successful surgeries, with over 3,00,000 surgeries being performed annually worldwide. A retrospection of the history of development of THR prostheses is proof of how judicious material choice can increase surgery success rates. While early attempts using metals (such as stainless steel) failed due to material design-related issues, the use of a Teflon™ acetabular cup by Sir John Charnley, a British orthopaedic surgeon, marked the beginning of the modern age in orthopaedics in 1961.

High-density polyethylene and stainless steel are currently used for THR and other orthopaedic applications. Ever since the first intraocular lens was discovered by Sir Harold Ridley in 1949, research and development in biomedical materials has grown exponentially with the advent of new materials, technological expertise, and novel applications. One such landmark is that of the bioglass developed by Larry L. Hench in the late 1960s.

In the 1960s to 1980s, many advances in the field of biomaterials were made and new materials were being explored, such as silicone, Teflon™, hydrogels, and bioglass. In 1982, the first artificial heart was placed in a patient as a temporary measure, and the patient lived for 64 hours. Subsequently, the Jarvik-7 model was permanently implanted into a human subject, who lived for 112 days. In 2006, Anthony Atala at Wake Forest Institute for Regenerative Medicine, USA, grew and implanted a tissue-engineered bladder in human patients.

By 2019, examples of successful translations of biomaterials research include hip and knee implants, dental implants, artificial kidney, breast implants, vascular grafts and stents, pacemakers, and heart valves, etc. The range of applications for these biomedical devices and their widespread use are commensurate with their ever-increasing numbers. Thus, it can be rightly stated that biomaterials have become an integral part of our lives today. Figure 1.2 shows the anatomical areas in the human skeletal system, wherein synthetic biomaterials can be used for repair/replacement of the injured part. In most of these cases, the materials, employed originally, were off-the-shelf materials, without considerable redesign and were developed primarily for

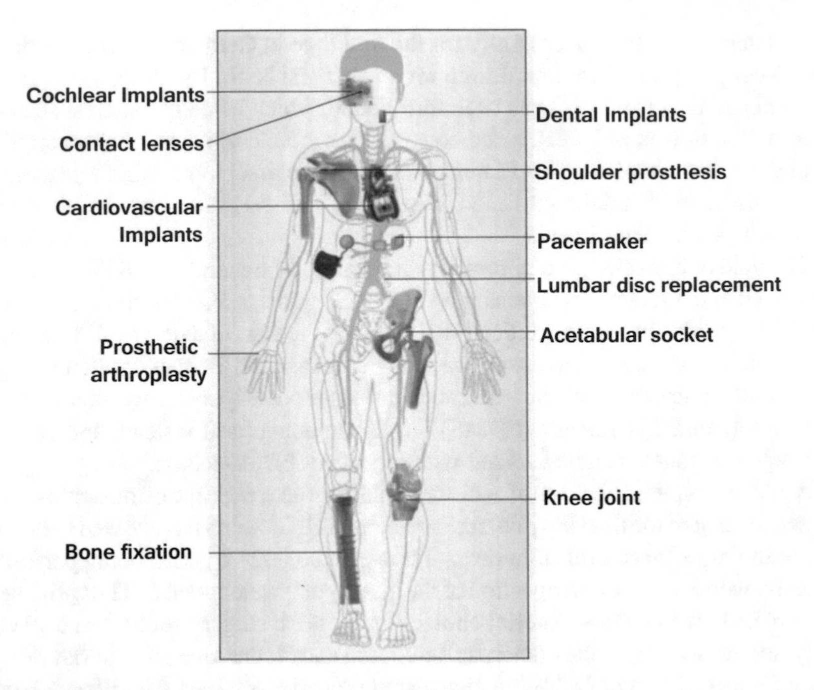

Fig. 1.2 The spectrum of clinical applications of biomaterials. (Adapted from Alijani et al. 2018)

non-biological applications. Their interaction with biological systems was not sufficiently understood until four decades ago. To this end, researchers have made significant efforts in the last two decades to create new materials with specific requirements for particular biological applications.

1.2 The New Science

A biomaterial is ideally destined to favourably interact with the components of the living system. The multi-cellular living organism has a highly structured, hierarchical organization, with each level having progressively increasing structural complexity. In order of increasing complexity, these levels are cell, tissue, organ, and organ system, respectively. A cell is the smallest unit of a biological system containing various cellular organelles. A tissue is a self-organised array of similar types of cells performing a similar function. At the next level, organs are formed when the identical types of tissues, having a common function, are grouped together. Various organ systems work together to maintain the life of a living organism.

The functionally related organs constitute a higher level of organisation, known as an organ system. Vertebrate animals have many organ systems. For instance, the digestive system includes organs, such as the oesophagus, intestine, liver, and gall bladder. In the human body, several thousands of cells of different types with millions of proteins co-exist together.

From a scientific standpoint, each cell type prefers to adhere, proliferate, or differentiate (change of functionality) on a biomaterial with specific composition, in a manner dependent on elastic stiffness, surface energy/surface wettability.

The difference in preference or sensitivity for various cell types towards different substrates substantiates the fact that the biomaterial composition or surface properties should be tailored for a specific biomedical application.

Let us use a conceptual tetrahedron to illustrate the new science of biomaterials to fabricate human tissues and organs (Fig. 1.3). The traditional material science tetrahedron emphasises the interrelation between the four main areas of a material: processing, microstructure, properties, and performance. To illustrate the development cycle of biomaterials, I have adapted the tetrahedron to encompass the key aspects of this process, as shown in Fig. 1.3. For biomaterials, two distinct manufacturing techniques are particularly relevant: first, additive manufacturing (for instance, 3D printing, 3D bioprinting, 3D plotting, etc.), that has positively influenced fabrication of patient-specific implants; second, porous scaffold manufacturing techniques (e.g. salt leaching, and gel casting) that are integrated with biological cells or molecules for regenerative medicine applications. The requirement of having porosity in materials for specific biomedical applications (e.g. tissue engineering) necessitates one to use a specific set of processing routes.

Concerning structure, the bulk microstructure and more importantly, surface, are extremely significant in the context of interaction with biological systems. In particular, grain size or phase assemblage of a material in bulk determine many macroscale

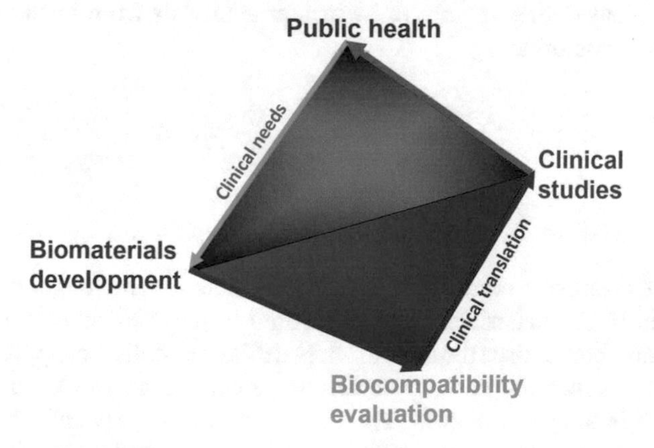

Fig. 1.3 Schematic illustration of the Biomaterials Science Tetrahedron, a conceptual framework used to take laboratory-scale research to patient's bedside in clinic

properties, like strength, toughness, etc. The surface topography, i.e. roughness or presence of specific patterns, guides the way in which the cells will function on a material surface. As part of the bulk structure, 3D porous architecture is particularly important in the context of bone tissue engineering. The fine scale characterisation, in terms of pore size distribution and interconnectivity in 3D space, requires one to use specific characterisation tools, e.g. micro-computed tomography, which are not widely used for various structural or functional materials. Similarly, reproducing the 3D architecture of the extracellular matrix is very crucial for scaffolds in soft tissue regeneration.

In the context of properties, the bulk strength or toughness, and more importantly, elastic modulus, of an implantable biomaterial play an important role in the biomechanical compatibility of a material in an osseous system.

In addition, the biocompatibility is to be sufficiently and ethically characterised using a host of biochemical assays, and molecular biology techniques, etc. This makes the field of biomaterials distinct from other fields of materials science. This latter aspect requires the researchers to develop an appropriate level of understanding of biological systems (cell, protein, bacteria, blood, etc.) and also to adopt techniques of remotely linked disciplines, e.g. biological sciences.

The performance of an *in vitro*-tested biomaterial is to be assessed in pre-clinical studies pre-approved by the ethical committee of the institute, involving experimental animals. The selection of animal model, defect type, as well as study duration or follow-up evaluation, depend on the specific biomedical application of a material. Such studies also involve expertise from veterinary surgeons, toxicologists, histopathologists, etc. The results of such studies provide scientific understanding of tissue-material interaction. Once a material is proven to be biocompatible in animal studies, the performance can be assessed in human clinical trials. It should be emphasised that clinical studies require even stricter approvals from institutional and/or national ethical committees.

Once the outcome measures, over post-surgical periods in human subjects are found to be clinically acceptable, a biomaterial implant or scaffold can be commercialised for use in humans.

The above discussion certainly indicates an interactive and longer product development cycle for biomaterials, when compared to those of other non-biomedical materials. It also emphasises the need to adopt a different skillset or approach to establish processing-structure–property-performance correlation in the case of biomaterials, which is unique in nature. The interaction among several remotely linked domains of expertise makes the field interesting and challenging (see Fig. 1.1). Therefore, biomaterials science encompasses a multidisciplinary approach: from engineering to biological to medical sciences (see Figs. 1.1 and 1.4).

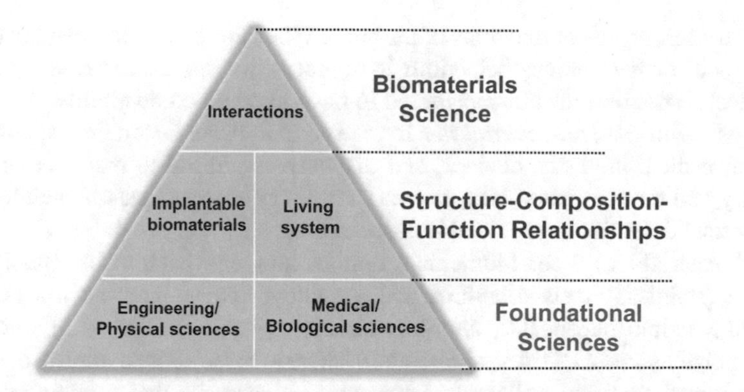

Fig. 1.4 Concept triangle to describe the field of biomaterials science (Adapted from: Biomaterials Science and Tissue Engineering: Principles and Methods, Cambridge University Press, 2017)

The students from non-biological sciences always seek to find the answer as to how far, or to what extent, they will need to learn or understand the core of the biological sciences discipline. While the fundamentals of engineering ideas can be translated to fabricate and characterise synthetic materials, an understanding of biological sciences is necessary to analyse the interaction of proteins, cells, bacteria, and blood with synthetic biomaterials.

The biocompatibility is studied both *in vitro* (laboratory-scale experiments in glassware or in culture/growth medium) or *in vivo* (study conducted in whole organism, i.e. animal models).

An integrated understanding of cells/tissue interaction with a biomaterial constitutes the central theme of the field of biomaterials science. Also, the interaction between these disciplines makes biomaterials science a truly interdisciplinary field of research (see Fig. 1.4). The discussion in the previous sections indicates that although biomaterials science is a relatively young field, it has made significant impact on human society.

1.2.1 The Key Elements

A biomedical material, or biomaterial, possesses many unique properties that allow it to interact with living systems: proteins, cells and bacteria (see Fig. 1.4). Many definitions have been proposed over the years for the terms 'biomaterial' and 'biocompatibility'. To begin with, biomaterials have been defined as natural or manmade materials that can interface with biological systems, to evaluate, treat, augment, or replace tissues, organs or functions of the body. When these materials display universally 'good' or harmonious behaviour in contact with tissues, evoking a minimal biological response; they are considered to have good biocompatibility. According to another definition, *biomaterial* is a term used to indicate materials that constitute parts of medical implants, devices, and disposables utilised in medicine, surgery, dentistry, and the veterinary field, as well as in every aspect of patient healthcare. A more formal definition is provided in Appendix A.

The materials used for biomedical applications can be broadly classified as metallic (stainless steels, titanium and its alloys, cobalt–chrome alloys, etc.), ceramic (calcium phosphates, alumina, zirconia, glass-ceramics, etc.), and polymeric (polyethylenes, PMMA, nylon, etc.). There are also some naturally occurring materials, such as fibrin, collagen, hyaluronic acid, and silk that are being used for biomedical applications.

It is important that a biomaterial must integrate or interact, in a clinically acceptable manner, with the components of the living system.

At this point, it is necessary to highlight how the concepts of biological sciences are adapted to biomaterials science. A biological cell can function in a different manner on a material substrate (underlying substance), when compared to that in a culture medium (in isolation). Although an understanding of cell fate processes (cell growth/proliferation, cell migration, etc.) from the biological science point of view is necessary, this needs to be adopted considering that a cell has to interact with a material substrate. Such interaction is mediated via adsorbed protein on the surface of a material. Similarly, the growth kinetics of bacteria in culture (in isolation) can be different as to how they multiply in a microenvironment around a biomaterial. This also has significant implication towards biofilm formation, which leads to prosthetic infection.

Another important idea from biological sciences is how to qualitatively and quantitatively analyse cell functionality modulation on a biomaterial. This aspect is often supported by how specific genes are upregulated/downregulated, which requires the quantification of expression of mRNA, which are extracted from cells grown on a material. With the advancement of the field in terms of gene-level analysis, biomaterials science is getting more and more firmly integrated with the biological science discipline. Once a biomaterial is screened to have acceptable or desired cytocompatibility, the next important step is to evaluate the tissue-level compatibility of a biomaterial, implanted in an animal model.

It needs to be emphasised that the central concept of biomaterials science intelligently uses necessary elements of the biological science discipline at an interface with materials science.

1.2.2 Defining Biocompatibility

One of the pre-requisites for a successful biomaterial is its acceptance by the host. This interaction and coexistence of a biomaterial, be it synthetic or natural, with the biological system is the fundamental meaning of biocompatibility (see Fig. 1.4). While it is relatively common to describe a successful biomaterial as biocompatible, the exact nature of biocompatibility is still uncertain. Along the lines of the first principle of Hippocrates that a doctor should do "no harm", the main requirement for biocompatibility of a material, whatever its end application, is that it should do "no harm". Biomaterials are distinct from other classes of materials in the sense that they have to be biocompatible with the biological system.

Phenomenologically, the biophysical events at the interface of a synthetic material and the biological system have been described in Fig. 1.5. Proteins are adsorbed within seconds of a material being placed in a biological system and their conformation changes can only be probed using molecular dynamics (MD) simulation or other computational studies. If the adsorbed proteins express desired biological functions, then they mediate the interactions with the biological cell. In an effort to establish better interactions, a cell changes its morphology and can expand on a material surface. Subsequently, cell–cell interactions and extracellular matrix formation take over in the chronology of the biophysical events. At the tissue level, ECM deposition

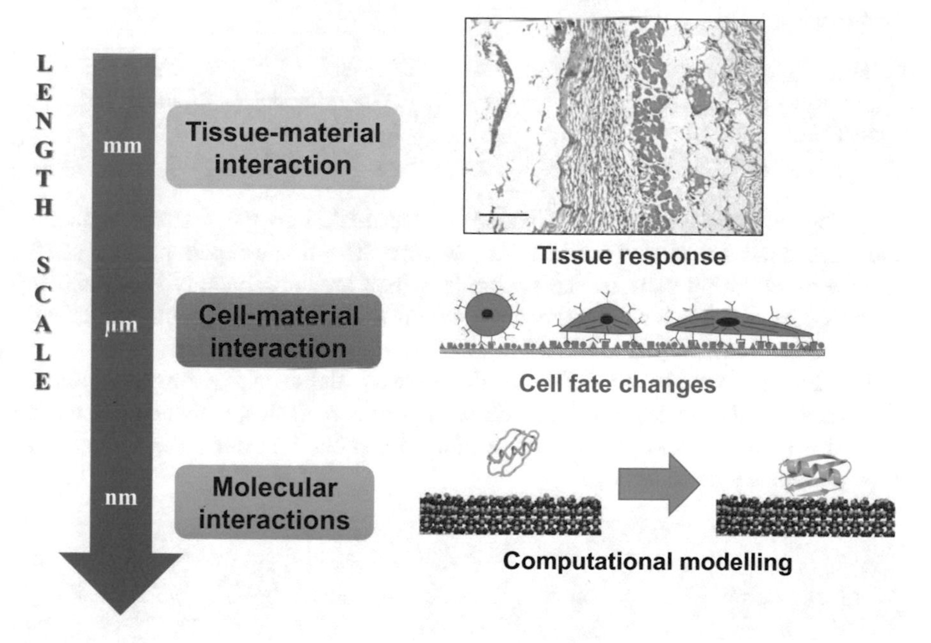

Fig. 1.5 Schematic of the interactions of an implantable biomaterial with the living system at different length scales

is followed by fibrous capsule formation in case of bioinert materials. In the case of biodegradable materials, tissue formation synchronises with scaffold degradation.

The concept 'biocompatibility' is to be defined in the context of the targeted clinical application.

It must be mentioned here that biocompatibility is application specific. This is primarily because, different types of cells have different surface preference, and accordingly, they adhere, proliferate, or differentiate. Such differences necessitate that material surface properties be tailored for a specific biomedical application. A multitude of issues—for instance, lack of desired physical properties (elastic modulus/toughness), piezoelectric property (the ability of certain materials to generate an electric charge in response to mechanical stress), and bactericidal property (killing bacteria)—have triggered significant research activities on biomaterials and implants.

Two key components need clarification. Biomaterials are directly linked to tissue engineering applications for the regeneration of functional human tissues. Although the body has its intrinsic self-healing properties, the need for healing and repair arises due to injury or disease. Biomaterials are designed to provide an architectural framework to encourage cell growth and tissue regeneration to enable biological functions.

Scaffolds are physical structures of synthetic or biological material, or both; that allow cells to attach, migrate, or modify their behaviour.

Biological cells are generally loaded (or 'seeded') into the scaffold to facilitate three-dimensional tissue formation, *in vitro*. The tissue repair process starts *in vitro* by culturing cells on the scaffolds, which are subsequently implanted in the host. Usually, this technique requires porous scaffolds to be biocompatible and bioresorbable (naturally dissolving).

It is difficult to mark a line of distinction between tissue engineering and biomaterials science. Tissue engineering deals with porous scaffolds, which can be made of ceramics, metals, and polymers as well as biological materials, like collagen or other proteinaceous materials.

Nonporous biomaterials, which are used for load-bearing applications (e.g. articulating joints) are commonly defined as implants.

Common implant examples are ceramic femoral heads and stainless steel or titanium-based stems in total hip joint replacement (THR) surgeries. The interactions of scaffolds or implants with biological systems are fundamentally different, and constitute the central theme of biomaterial science.

1.2.3 Structural Concepts

There are four kinds of biological macromolecules: carbohydrates, lipids, proteins, and nucleic acids. Macromolecules such as polymers, are composed of many similar smaller molecules linked together in a chain-like fashion. The smaller molecules are called monomers. The most important macromolecule in a living system is the protein and its monomer is the amino acid. An important example of a protein is collagen, which is the most abundant protein in the human body. Proteins constitute more than 50% of all major intracellular organic molecules and have a diverse range of functions in a cell. Each cell in a living system may contain thousands of different proteins.

In general, amino acids are structurally characterised by a carbon atom (the alpha carbon) bonded to the four groups—a hydrogen atom (H), a carboxyl group (-COOH), an amino group ($-NH_2$) and a 'variable' group or 'R' group. The reaction of two amino acids forms a peptide bond. The 'R' group varies among amino acids and determines the differences among protein monomers. The structure of a protein is specified at different levels, namely as primary, secondary, tertiary, and quaternary structure. Secondary, tertiary, and quaternary structures of proteins can be differentiated from one another on the basis of their degree of complexity.

The cells are the smallest structural and functional units of life that make up all the plants, animals and single-celled self-sustaining organisms, such as bacteria and yeast. In general, eukaryotic ('truly' nucleated) cells are enclosed by a lipid bilayer and contain the necessary genetic material, needed to direct the continued propagation of the cell. All the living cells, whether unicellular micro-organisms or just a tiny part of a multi-cellular organism, look widely different, but share certain characteristics in common. They are small 'pockets', composed mostly of cytoplasm (consisting of cellular machinery and structural elements, such as free amino acids, proteins, carbohydrates, fats, and numerous other molecules) enclosed within a phospholipid bilayer membrane.

As far as different cell types are concerned, adipocytes help in storing/synthesising fat compounds and constitute adipose tissues. While neurons are important cells of central and peripheral nervous systems, astrocytes are a type of cell providing structural support to neurons. In bone tissue, three types of cells are present. First, osteoblasts, which are known as bone-forming cells, and second, the osteoclasts which are bone-resorbing cells. Third, osteocytes are the matured osteoblasts. All the three bone cell types are morphologically distinct. In skin tissue or endothelium, the endothelial cells with close spacing are present and they protect the underlying cellular structure.

Tissues are at an intermediate level between cells and a complete organ in the hierarchy. Four classes of tissues are connective, muscle, nervous, and epithelial tissues. Connective tissues have fibrous morphology to provide structural support to organs. Among different types of muscle tissues, smooth muscle is found in the inner linings of organs, while skeletal muscle is attached to bone. Cardiac muscles enable the heart to contract and pump blood throughout the organism. In the central nervous system, neural tissue is contained in the brain and spinal cord, while in the peripheral nervous system, the cranial and spinal nerves consist of neural tissue, which contains neurons or nerve cells. Epithelial tissues contain closely placed epithelial cells, which line organ surfaces.

Tissues can be classified into soft and hard tissues, primarily based on the elastic stiffness or mechanical strength property. Hard tissues are mechanically superior to soft tissues, and some of these tissues have desirable adaptability and multifunctional properties. Mineralised tissues (bone, tendon, cartilage, tooth enamel and dentin) are examples of hard tissues, which contain hard minerals and soft collagenous matrices.

> Despite extensive research on synthetic biomaterials, often via biomimicking approaches, researchers are yet to develop a synthetic material with matching mineralised tissue properties, and therefore, newer biomaterials design approaches are under development.

1.3 Stem Cells and Regenerative Engineering

In view of its central importance in biomaterials science and biomedical engineering, this section discusses stem cells and their importance in regenerative engineering. Over the last one decade, stem cells and their promising applications in healing and regenerative therapies have been the central foundation for a new scientific discipline, **regenerative engineering**. Stem cells can differentiate into various cell types and are fundamentally different from other cell types in terms of the following characteristics: (a) Self-renewal: The ability to divide, usually after long periods of time, into more stem cells while maintaining their undifferentiated state. (b) Potency: The ability to differentiate into specialised cell types. It is known that stem cells are of five types; these are (a) totipotent: can differentiate into any kind of cell of a total organism, (b) pluripotent: can differentiate into all cell types, (c) multipotent: can differentiate into a number of closely related cell types, (d) oligopotent: can differentiate into only two cell types and (e) unipotent: can differentiate into a single or specific cell type; (e) clonality, the ability to be derived along a specific lineage or from one source. In cell biology, a clone represents a group of identical cells sharing a common ancestry.

Stem cells can be derived from different sources, and accordingly, two major classes of stem cells include, (a) embryonic stem cells (ESCs)—obtained from the human embryo during the process of *in vitro* fertilisation. ESCs differentiate into all the cell types and therefore are totipotent in nature. (b) Adult/somatic stem cells—found in the differentiated tissues or organs, and these can renew themselves and differentiate into major cell types (pluripotent in nature). They are mainly involved in repair and maintenance of tissues or organs.

Further, stem cells can be classified into three types, (a) haematopoetic stem cells, which can differentiate into blood cells, (b) mesenchymal stem cells, which are derived from bone marrow and can differentiate into multiple cell types, and (c) adipose stem cells, which are derived from fat cells. There are two other stem cell types; (i) amniotic stem cells found in the amniotic fluid and (ii) induced pluripotent stem cells (iPSCs), wherein the former type can be sourced from the amniotic fluid. iPSCs are mature cells genetically reprogrammed by certain transcription factors that have a pluripotent stem cell state.

The seminal work by Takahari and Yamanaka on the discovery of induced pluripotency was awarded the 2012 Nobel Prize in Physiology and Medicine; jointly with Sir John B. Gurdon.

In the field of medical science, significant clinical research is currently underway to use stem cells (stem cell therapy) for the treatment of the host of diseases, such as diabetes, rheumatoid arthritis, Parkinson's disease, Alzheimer's disease, osteoarthritis, repairing hearing and vision loss, cardiac infarction, Crohn's disease and also for wound healing applications.

The impact of stem cells in tissue engineering can be substantiated with the help of a few specific examples. Cartilage tissue contains the characteristic chondrocytes and the differentiation of hMScs to chondrocytes is known as chondrogenesis. The myogenesis of hMScs leads to the formation of smooth muscle cells, which are present in the musculoskeletal system and also in other parts, e.g. lining of blood vessels or digestive tract. Similarly, cardiomyogenesis of hMScs results in the cardiomyocytes, which form the cardiac tissue and this cell type is one of the known cell types, which does not grow easily in culture. In patients suffering from blood cancer, one of the clinical treatment options can be to transfuse haematopoietic cells. Another important cell type is pancreatic islet cells, which has larger relevance in controlling blood sugar level and is therefore relevant for the treatment of diabetes.

1.4 Unmet Clinical Needs and Clinical Perspective on Biomaterial Implants

In the last few decades, human healthcare needs have stimulated research activities in the fields of biomaterials science and biomedical engineering. Studies indicating significant evolution in materials science and engineering in terms of developing potential biomaterials for healthcare have been widely published in the literature during the last few decades. Commensurate with improvements in medical science, considerable research has been pursued in the areas of biological, chemical, and physical sciences. In this perspective, researchers made considerable efforts to understand the molecular biology aspects of some of the life-threatening diseases, and engineers have attempted to develop various diagnostic tools, as well as synthetic implant materials, to assess the disease state or to repair the damaged tissues. The quantification aspect in the disease diagnosis has enabled better healthcare or more timely treatment.

Among various diseases impacting human life, cardiac-related diseases, cancer and neurodegenerative diseases are among the life-threatening diseases, while a large number of patients also suffer from orthopaedic, i.e. bone/joint-related diseases. The translational research on biomaterials should, in principle, therefore be driven by the unmet clinical challenges in the global scenario.

In most cases, complicated revision surgery is needed after a few years of implantation. The discovery and design of a material for hard tissue replacement with the desired characteristics, namely long-term durability without significant degradation in physical and mechanical properties, is a major area of research in orthopaedics. In different anatomical parts of the human patient, metals, ceramics, polymers, and their composites with acceptable biocompatibility are being used as artificial implants.

In vitro biocompatibility assessment requires expertise from molecular biology, microbiology, and other branches of biological sciences. An *in vivo* biocompatibility study requires the participation of a veterinary surgeon, to assist in experiments with small, medium to large-sized animals. The safety evaluation of biomaterials requires toxicity studies, both at cell and gene levels. This requires the involvement of toxicologists.

A number of examples related to human diseases will be provided below to substantiate the above discussion. For example, a femoral head in total hip joint replacement (THR) needs to sustain complex biomechanical stresses, which are to be evaluated using burst strength measurement and hip joint simulator experiments. Similarly, the heart valve prototype has to sustain a large number of fatigue cycles. Biomechanical considerations are equally important in dental restorative applications. The subsequent sections highlight key clinical needs in India, with reference to some of the developed nations, however, this is not an exhaustive list.

1.4.1 *Musculoskeletal Surgery and Orthopaedics*

Annually, around 1,20,000 knee replacements are carried out and 75,000 hip replacement are carried out in India. The American joint replacement registry (AJRR) has reported an exponential growth in the number of total hip replacement (THR) procedures in the USA during the last decade. In addition, the burden of revision surgery is more than 10%. In an increasingly ageing—and obese—world, weight-bearing joints of the hips, knees, and spine are under siege. Total hip or knee replacements are rising exponentially. It is estimated that by 2030, the number of total knee replacements performed in the USA will increase by more than 600% compared to that of 2005, while total hip replacements are expected to increase by almost 200% over the same time period. In India, knee arthritis is emerging as the fourth most common cause of physical disability, while the joint replacement market is growing in double digits.

Among the musculoskeletal diseases, arthritis or other rheumatic diseases are most prevalent in India.

Clinically, two variants of arthritis are osteoarthritis (OA) and rheumatoid arthritis (RA). These diseases are associated with severe pain and in the most critical conditions, can potentially render the joints dysfunctional.

Many orthopaedic surgeons regularly treat osteoarthritic patients, who visit the clinics with the complaints of severe pain around all types of load-bearing joints like hip, knee, and spine joints. X-ray radiography analysis mostly reveals the wear and tear of joints as the primary cause of such problems, often affecting the cartilage between two joints, or other problems, like bone overgrowth and joint damage.

Clinical investigations over last few decades reveal the concept of aseptic loosening, initiated by the release of significant wear debris particles from the articulating joint surface (see Fig. 1.6). These wear debris particles can cause downstream inflammatory reactions, leading to the activation of osteoclasts, leading to resorption of the bone cells. RA normally is initiated at the linings of load-bearing joints and is the manifestation of the attack by body's own immune system. RA can be even more painful as it may affect other joints and sometimes organs, like heart or lungs. Apart from OA or RA, osteoporosis is another musculoskeletal disease, involving thinning and weakening of bone.

Osteoporosis is commonly attributed to bone loss and deficiency of the bone mineral content in elderly patients.

In the extreme case of severe functional loss of joints, surgical intervention by the orthopaedic surgeons is the only treatment option. The natural joints in the musculoskeletal system can be replaced by the synthetic implant system in orthopaedic

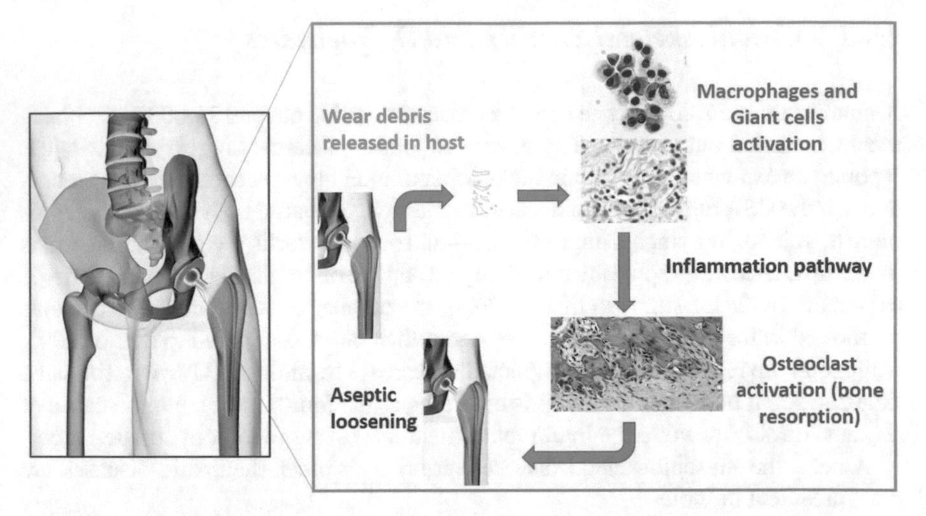

Fig. 1.6 Clinical perspective of total hip joint replacement (THR): Femoral head (Co-Cr) and Acetabular socket (UHMWPE) in THR surgery—such devices are reported to release wear debris particles, causing inflammation (need for revision surgery)

surgeries to restore natural joint-like functionalities in the treated patients, and generically, this treatment option is known as total joint arthroplasty (TJA). These clinical procedures are commonly named after the type of joint that is being replaced. For example, the clinical terms, total hip joint arthroplasty (THA) refer to the replacement of the hip joints; similarly, total knee arthroplasty (TKA) refers to knee replacement.

In the case of THR, commercially available femoral hip stems are an example of an ongoing innovation (Fig. 1.6). The material for the femoral stem can be typically stainless steel or Ti-alloys. The acetabular liner-femoral head combination can be ceramic–polymer or metal–polymer or metal–metal. Among these options, ceramic–polymer mating combinations are by far the most widely accepted clinically, because of their better biocompatibility property. As far as THR assembly integration is concerned, the press fit concept, as well as cementless or cemented hip systems, are the two major options. While bone cement (e.g. PMMA) is used for the cemented stem, this is not the case for the cementless stem, wherein bioactive HA is coated onto the stem for better osseointegration.

For the problem of osteoarthritis, which is associated with chronic joint pain, the potential solution is an implant with improved wear resistance and mechanical strength. The development of a chondrogenic scaffold and injectable cartilage are proposed as potential solutions to osteoarthritis. Osteomyelitis is associated with infection at the fracture site of bone which might spread to the nearby tissue.

Biocomposite mixed with antibiotics (STIMULAN®) is already in the Indian market, but it is still not affordable for a large cross-section of society.

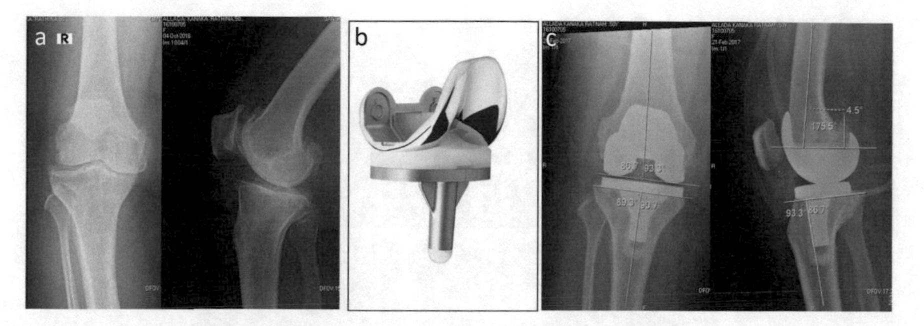

Fig. 1.7 (**a**) Clinical radiograph of osteoarthritic knee, (**b**) Total knee replacement implant, (**c**) Post-operative radiography after knee surgery

In terms of patient specificity, different sizes of femoral head and respective liner combinations are commercially available. In the case of knee replacement surgery, the device components have different sizes and shapes than hip replacement. In particular, the knee femoral component and tibial trays are the major components of TKR (Fig. 1.7). In cases of accidents or trauma, the reunion at the site of bone fracture and osseointegration (integration of non-biological origin implant to an osseous system) are some of the clinical challenges. There have been recent advances in the use of biodegradable polymers to address such challenges.

1.4.2 Dental Restoration

The poor state of the nation's teeth is evident from reports that reveal that over half of Indians have experienced tooth loss. Tooth decay has become an epidemic.

The level of complete edentulism (toothlessness) in the general population varies from 14 to 16%.

Edentulism in the elderly, above age 65, is around 91.25% in both urban and rural India. An initial cost estimate analysis indicates that indigenous dental implants are twice as cheap when compared to the cost of imported implants in the market.

With respect to dental reconstruction, a few clinical cases highlighting the use of implants in dental surgeries are shown in Fig. 1.8. Abfraction is a physico-mechanical loss of tooth architecture, that is not caused by tooth decay and is located along the gum line. This affects mostly the enamel and the dentin part of the crown. The development of moderate and steady peptide-releasing biomaterials, which will be able to regenerate or cure the area, as well as a temporary supporting material in the affected region may be a probable solution. Another challenge is to address dental caries, which can be described as cracks, channels, pits, grooves, and cavities in

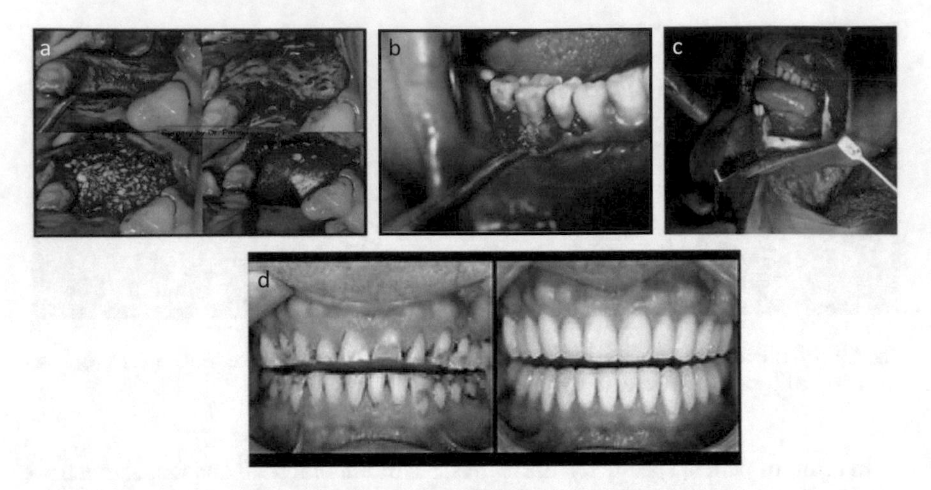

Fig. 1.8 Dental surgeries requiring implants (**a**) Ridge augmentation (**b**) Peridontal surgery with grafts (**c**) Mandible surgery (**d**) Full mouth rehabilitation

tooth structure. This can originate in the back teeth, between teeth, around dental filling or bridgework and/or near the gingival margin. The development of a suitable biomaterial, which will be able to form a strong and long-term bond with dentin for treatment of caries, can be a potential solution. Edentulism requires the use of dental implants, which can be either metallic (e.g. Ti6Al4V) or ceramic (e.g. ZrO_2). The implant is placed into the jawbone of the patient, and on the top of an implant, the dental crown is to be fitted to restore the mastication (i.e. chewing) function in a patient.

The dental implants have specific geometric contour and designed thread profiles for better post-operative stability and osseointegration.

1.4.3 Neurosurgical Treatment

In the context of neurosurgical applications, it is important to design biomaterials that improve interaction among neurons, bone, and meninges of the nervous system. Also, prevention of scar tissue formation by astrocytes is extremely important for neurological/cranial-based applications. Three major unexplored challenges in the Indian context are childhood skull deformation, adult skull deformation, and duraplasty. For the first two applications, the problems associated with titanium implants need to be addressed. For duraplasty, the polypropylene and collagen-based implants can provide potential biomaterials-based solutions.

Among various neurosurgical procedures, decompressive craniectomy (DC) is one of the commonest treatment options in the management of traumatic brain injury or stroke.

DC is a neurosurgical procedure performed in clinical cases of raised intracranial pressure and impending herniation secondary to TBI or stroke. In such procedures, a pre-defined part of the skull is removed. This first step allows the raised intracranial pressure to return to the normal level and prevent any brainstem dysfunction. In order to restore the desired functioning of the underlying brain cortex, the craniectomy defect is to be closed with autologous bone or an implant (bone flap), in a follow-up surgery known as cranioplasty. A large array of materials ranging from auto/allografts and synthetic materials such as acrylic, ceramics, and polyether ether ketone (PEEK), with or without osteo-inductive growth factors have been used by neurosurgeons to repair cranial bone defects.

In cranioplasty surgery, titanium and stainless steel-based alloys have been established as the most promising implant materials, but other materials such as PEEK or PMMA are also clinically used with reasonable success in terms of desired functional outcome (an example is shown in Fig. 1.9). Biodegradable polymers are most widely used as screws and cranial plates; however, their fracture during implantation is a disadvantage. All these clinical and associated issues require focused translational research programmes in this direction.

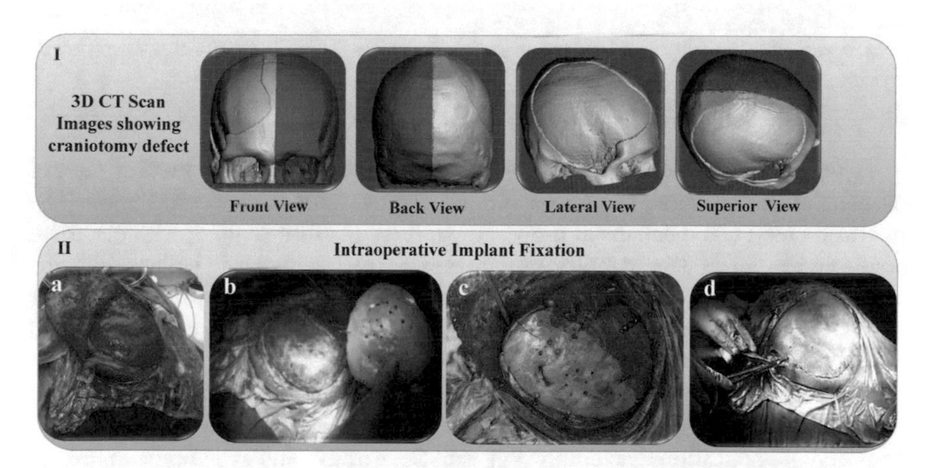

Fig. 1.9 A recent clinical study on cranioplasty surgery by neurosurgeons in collaboration with the author's group. I Conversion of CT Scan DICOM data to stl* file by 3D Slicer for modelling of the craniotomy defect. II Intraoperative cranioplast implant fixation. (**a**) Depressed area with incision: skin flap and pericranium are retracted exposing the dura (**b**) The dura is incised relieving the subdural pressure (**c**) Polymethyl methacrylate (PMMA) custom bone flap is fixed with titanium screws/plates (**d**) An epidural drain is placed and the skin is closed with staples

In many neurosurgical treatments, the current clinical therapeutic methods for the rescue and repair of damaged neural tissue are reported to be ineffective, as they do not provide functional recovery of the nervous system.

Several research groups in India and abroad have worked towards the use of functional biomaterials-based platforms, and further combining them with stem cell-based therapies to mimic the natural microenvironment of the glial, neuronal, and stromal cells of the nervous system.

The functionalities of biomaterial-based platforms provide 3D matrices with desired pore sizes, porosities, elasticities, and wettability along with various chemical, electrical, biological, and topographical cues that favor cellular attachment, growth, proliferation, directed alignment, and differentiation as well as proper nutrient flow for neural tissue regeneration.

Furthermore, considering the inherent presence of electrical fields and synapses in the nervous system, the conductive biomaterials-based platforms, such as films, hydrogels, fibers, composites, and flexible electronic interfaces, biophysical stimulation have also been used through application of electrical stimuli to enhance the nerve regeneration process. These platforms have been particularly used for controlling neurite extension, directed migration of neuronal and glial cells, and differentiation of stem cells.

The conductive polymer or carbon-based platforms hold significant potential to manipulate cellular behavior and to promote neural regeneration.

However, the properties of these platforms, along with their precise control on cellular mechanisms, should be further improved for clinical translation. The microstructural and mechanical properties of these platforms, along with electrical conductivity and biodegradability, require certain improvements to better mimic the extracellular matrix (ECM) and to be applicable in clinic. There are still a number of unknown aspects regarding the relationship between the electrical properties and cellular functions, particularly from a mechanistic perspective.

Elucidating these relationships will pave the way for development of novel and functional platforms not only for neural tissue engineering, but also for other tissue engineering applications including skin, muscle, cardiac, and even brain–computer interfaces. In addition, the use of biodegradable conductive platforms in tissue engineering applications can potentially enable minimally invasive implantation and manipulation of cells through remotely applied electrical stimuli. This can prevent additional surgical interventions, therefore improving the experience of patients.

Cervical arthroplasty after anterior decompression with insertion of a prosthetic disc has been suggested as an alternative to anterior cervical fusion. Cervical disc replacement aims to maintain intervertebral movement. Currently, four prostheses with varying design principles (materials, range of motion, insertion technique and constraint) are available. Early studies show that in the short term, the complication rate and efficacy are no worse than for fusion surgery. As spinal surgeons enter a new era of the management of cervical spine disease, it is hoped that improvements in these prostheses are made in terms of implantable biomaterials, design, and clinical outcomes.

1.4.4 Cardiovascular Treatment

Cardiovascular diseases are considered as the primary causes responsible for increase in mortality rate worldwide. It has been recognised that scarred cardiac muscle results in heart failure for millions of heart attack survivors worldwide. When a blood vessel is blocked, the myocytes die due to oxygen deprivation, resulting in scarred tissue formation. The scarred cardiac muscle can potentially lead to heart failure due to gradual ventricular remodeling, unless the damaged area can be restored or replaced with new tissue or the blocked arteries can be decoagulated to remove the plaques using a stent.

Despite significant progress in the field of medical sciences, heart attack (myocardial infarction) is still regarded as a major killer even today. This is primarily because of the fact that it is difficult to regenerate the diseased cardiac tissues. In the case of myocardial infarction, i.e. 'heart attack', cardiac tissues are largely damaged.

In the heart, the functionality of cardiomyocytes (a contractile muscle cell that generates part of the myocardium tissue of the heart) and neurons (an electrically excitable cell that processes and transmits information by electrical and chemical signaling) depends on the continuous conductivity property.

However, such conductivity may break down during heart disease or malfunction.

A myocardial infarction can occur when a major blood vessel supplying the heart's left ventricle is suddenly blocked by an obstruction, mainly because of a blood clot (Fig. 1.10). This destroys the cardiomyocytes and neurons, leading to the formation of dead tissue, which ultimately results in myocardium denervation. Nerve damage to cardiac tissue can result in nerve growth in the left ventricle and development of arrhythmias.

The existing clinical treatment options include coronary artery bypass surgery, angioplasty, tissue-engineered cardiac patch, and heart transplant. In the case of angioplasty, coronary stents are used, and those expand when inserted into the damaged artery to clean up the undesired plaques. This results in better blood circulation in cardiovascular system. In case of the cardiac patch, the tissue-engineered scaffolds should have cardiac tissue-like conductivity property and the ability to

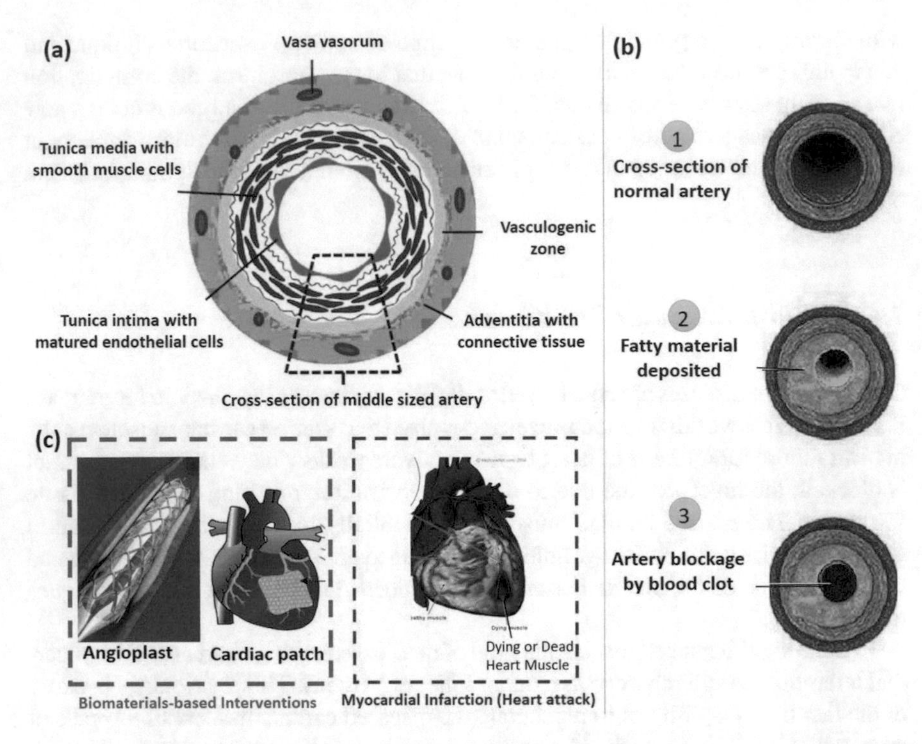

Fig. 1.10 Biomaterials based approaches for cardiovascular treatment. (**a**) Cross section of a middle sized artery. Adaptation of image by Bernhard B. Singer, Stem Cell Reviews (2008), 4 (3):169–77. (**b**) Cross section of an artery subjected to atherosclerosis. Adapted from an image by Carla Masia et al. Universita Degli Studi Di Pavia Thesis, "Constitutive modelling of a biodegradable polymer" (2007–2008). (**c**) Myocardial infraction and representative illustration of biomaterials based interventions. Angioplasty image source: www.medgadget.com

support the proliferation of the beating cardiomyocytes. Apart from the biocompatibility properties, it is important that that an ideal cardiac patch should have sufficient electrical and micro-mechanical properties to ensure the natural contractile nature of myocardium. The heart transplant is the chosen clinical treatment option, only in the event of total heart failure and the lack of the feasibility of the above-mentioned clinical treatment options. Among all these treatment options, the bypass surgery and heart transplant are very invasive in nature, while the cardiac patch-based treatment is minimally invasive in nature, with the possibility of fixing the dead cardiac tissue.

After implantation, a medical device or prosthesis undergoes a lot of modification at its surface due to various immune cellular complement activation, resulting from local tissue injury.

In cardiovascular physiology, a healthy endothelium not only plays an important role in controlling blood fluidity, but also inhibits the adherence and aggregation of macrophages and platelets.

Further, healthy endothelial cells (inner lining of blood/lymphatic vessels as well as inner lumen of heart) suppress the proliferation of smooth muscle cells (SMC) and regulate various inflammation processes. Therefore, the compatibility of blood contacting devices with endothelial cells and SMCs is important.

Five approaches are investigated in the bioengineering community for developing or improving the treatment options for cardiovascular diseases. One of the approaches is to deliver the stem cells to the infarct or direct injection; while in another approach, the cardiomyocytes from the patient's own pericardium are grown or printed by 3D bioprinting in a thin-layered form on a patterned hydrogel, followed by surgical implantation. Another tissue engineering-based approach is to deliver polymeric scaffolds with cell suspensions via injection or catheter. Alternatively, cells-seeded biological/synthesis 3D scaffolds are cultured in a 3D bioreactor and subsequently implanted into the infarct. The use of blood-compatible coatings on coronary stents is also investigated in an effort to develop better haemocompatible devices as well as to assess the ability to release blood-thinning drugs, when implanted into the diseased aorta. Another approach is the use of biophysical stimulation to regenerate the matured cardiomyocytes, often using a tissue-engineered scaffold or implantable biomaterials, as discussed in Chapter 3. Many of these promising approaches are still in pre-clinical validation or in clinical trial stage.

In the context of the bioceramic coated implants for cardiovascular treatment, it is noteworthy to mention that while Ti-based or NiTi-based shape memory alloys are clinically used as coronary stents, many of these materials do not have the greatest combination of blood compatibility and cytocompatibility. The use of bioceramic coatings, like TiN/TiAlN on Ti6Al4V, has been investigated and their blood compatibility has been established in various research groups. Apart from antibacterial properties and cytocompatibility with endothelial cells, the blood-contacting devices, like coronary stents, should have lower haemolysis, less platelet adhesion and aggregation, without causing any morphological deformation to red blood cells and extended blood coagulation time. In a recent research work from the author's group, a new generation of $SiC_xN_yO_z$ coatings on Ti substrate was developed for the cardiovascular application. These coatings demonstrated clinically acceptable haemocompatibility parameters and much better biocompatibility than uncoated Ti. These coatings have been further taken for pre-clinical validation.

Considering another treatment approach of electrical stimulation, it is important to recognise that the cardiac muscle tissue in the native heart has specialised electrical signal conduction and propagation pathways, such as electromechanical coupling via gap junction channels, and this coordinated electrical signaling mechanism is involved in heart contraction.

Cardiac pacing and defibrillation are among the most important heart disease treatment functions of electrical stimulation.

Cardiac arrhythmias or irregular heartbeat, leading to speeding up or slowing down of the heart rate, are treated by surgical implantation of a cardiac pacemaker in the chest close to the heart. An increasing number of publications showed the association of exogenous EF in inducing cardiogenesis in embryoid bodies (EBs) derived from hESCs. One of the greater challenges is to produce functionally mature cardiomyocytes with natural heart-muscle-like beating behavior.

1.4.5 Urological Treatment

In the field of urological treatment, clinicians need to treat patients with persistent medical conditions, that generally compromise the natural functioning of the urinary bladder to store and evacuate (void) urine (see Fig. 1.11). A large number of patients worldwide, estimated to be more than 400 million, are reported to suffer from urinary bladder-associated physiological disorders, with bladder carcinoma (BC) being the most common cause of morbidity. In fact, BC is considered to rank ninth globally,

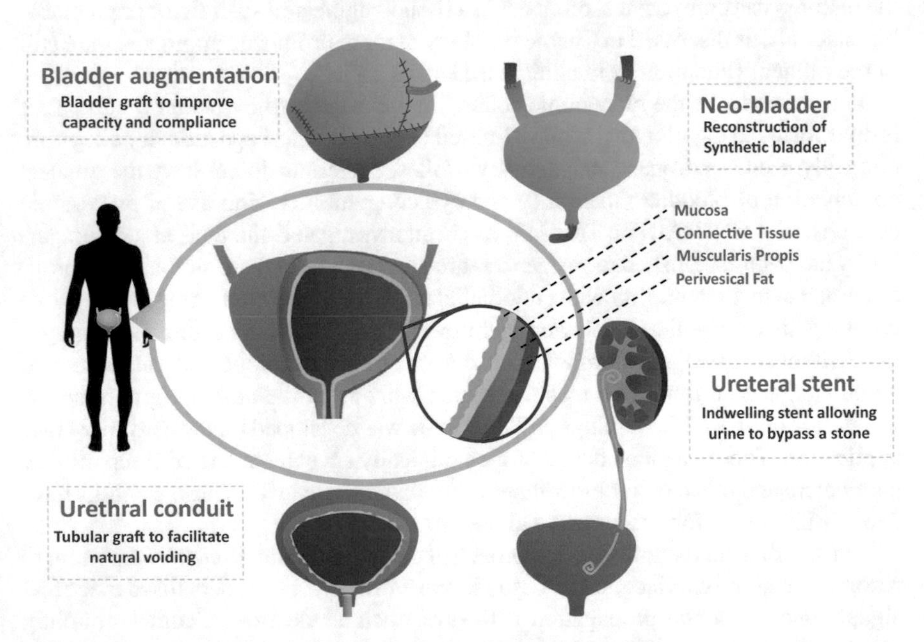

Fig. 1.11 Urinary bladder-associated pathologies and potential clinical applications of biomaterials in urological reconstructive surgery

with respect to other malignant diseases and is also the thirteenth commonest cause of cancer related deaths.

In India, three men and one woman, among every 1,00,000 citizens, develop bladder carcinoma every year.

To address these urological problems, one of the clinical treatment approaches is to design and implant a functional equivalent of natural counterpart, based upon the concept of tissue engineering and regenerative medicine. Another treatment approach is based on an alloplastic prosthetic bladder with a passive reservoir, which facilitates low pressure urine storage as well as subsequent volitional voiding. For the former approach, biodegradable polymers are to be used, and for the latter, the use of biostable/bioinert implants is preferred.

There is a current need for a new material for replacing the bladder in bladder transplantation. The neo-bladder should have desired properties and most importantly should be resistant to encrustation. Thus far, a pouch derived from the intestine has been used for bladder replacement. For surgery related to the urethra, urologists require a biodegradable material for stents and new biomaterials to replace the urethra and ureter, which do not get encrusted.

Concerning biomaterials in reconstructive urology, polyesters of alpha-hydroxy acids, such as polylactic acid (PLA), polyglycolic acid (PGA) and their copolymer, polylactic-co-glycolic acid (PLGA) are the key examples of synthetic biodegradable polymers to be developed into an implantable neo-bladder prototype and clinically tested in human patients. In recent times, a new approach of using hybrid scaffolds with the binary combination of a natural biopolymer and a synthetic biomaterial has been conceived in laboratory-scale research. In such an approach, the natural biopolymer provides the molecular signals to direct cell fate and cellular functionality, while the synthetic counterpart contributes towards structural reliability and tunable degradation.

The most successful clinical trials in bladder tissue engineering used the cell seeded hybrid scaffold of collagen and PGA, and this was conducted by Atala's group at Wake Forest Institute of Regenerative Medicine, USA.

Although scaffolds based on combinations of collagen with PCL or PLGA have been also investigated for potential urological applications, those have not yet been translated to human clinical studies.

1.4.6 ENT Surgery

The National Sample Survey Organization (NSSO), India, reported hearing loss as a major cause of speech disability in about 3 per 1000 persons with speech disability. Various reports from around the world have suggested that at least one child per 1000 is born with bilateral sensorineural hearing loss (SNHL) of at least 40 dB and four in 10,000 have profound SNHL.

As per World Health Organization (WHO) estimates, 5% of world population or 360 million persons have hearing loss, out of which, 28 million are children.

As per census bureau reports in 2010, 56.7 million out of 308.7 million people in USA have been living with some form of disability, corresponding to 19% of the total population, and among them, 10.8% of those with disability were noted to be below the poverty line. Over 7.6 million people experienced difficulty in hearing, including 1.1 million whose difficulty was severe. About 5.6 million people used a hearing aid.

Nearly, 27 million people are suffering from hearing disability in India. Among them, 5 million people have hearing disability amounting to 19% of total disability. The number of persons with hearing disability stands at 18.9% of 2.21% of total disabled population. Hearing disability is more among females compared to males and more in urban areas compared to rural areas. Among the total disability, the number of people having hearing disability increases with age and has been seen to reach its maximum level at 90 years of age. It has been reported that children account for 8% of the hearing disabled, adding up to 28 million among 5% of world population with hearing disability, although 7% of the hearing disabled people have congenital defects. During the rehabilitation process of hearing disability, it is required to retain the residual hearing of the patient.

The sensorineural-type deafness may require complicated surgical procedures including cochlear implant (CI) fixation. In India, all high-quality CI are imported, attributing to higher costs. The known clinical complications of cochlear implants generally include those related to surgical technique, post-operative care and device malfunction. Lesser known complications include facial nerve stimulation and non-responsive auditory nerve, which can be avoided. Though the success of CI is well documented and patient compliance is very high, CI can induce severe inflammatory reactions in many patients, which can cause damage to left over undamaged hair cells which will result in loss of residual low frequency hearing.

One of the major challenges in cochlear implant surgery is to preserve functional hair cells. Initial attempts are focused on a more refined surgical technique, though electrode modification is another option. The most important pathological pathway responsible for apoptosis of hair cells is excessive reactive oxygen species (ROS) production after CI fixation.

The electrode surface modification in cochlear implants to reduce ROS production in the auditory system represents one of the unmet clinical needs in the field of ENT surgery.

In this direction, a research programme in India is being taken up to find a solution to retain the residual hearing of hearing disabled patients, which is discussed in Chapter 3.

Limited resources, expensive devices, and the problems related to surgical techniques are major concerns for the success of the biomedical devices used to overcome hearing disability. As the biomaterial is an outcome of interdisciplinary effort from engineer, physician and biologist, a joint effort by all of them is expected to make successful devices and find suitable solutions for the existing problems of different biomedical devices.

1.5 The Tailpiece

Last but not least come all the concepts and elements that prop up biomaterials science. It is expected that all the concepts and unmet clinical needs, discussed in this chapter, should be used as guidelines to design biomaterials. The field is now, more than ever, a coming together of various science and engineering disciplines. It has seen the emergence of novel materials that have found critical medical applications: from bone replacement/healing to neurosurgical and urological applications. And at its heart lies the science, where biomaterials interact with living cells and tissues. However, the biomaterials of the future are expected to go far beyond that, where cutting-edge designs, biocompatibility, and computational insights are expected to provide better clinical solutions.

Chapter 2
Economic Impact, Healthcare Initiatives, and Research Funding

Abstract The potential economic and societal impact have driven significant research programmes on biomaterials and implants over the last few decades globally. However, the outcome has not yet been too significant in India, as more than 80% of the implants used in Indian hospitals are still imported! It is therefore important to consider the market value of implants and the quantum of business related to the field of biomaterials. In this context, this chapter introduces the potential economic impact of the indigenous development of high-quality biomedical implants, which also drives many national policies to facilitate translational research. Against this backdrop, governmental funding agencies have started several new initiatives. This chapter discusses the above aspects and briefly mentions the international status also. In addition, the present status of medical and bioengineering education is briefly mentioned.

B. Basu, *Biomaterials Science and Implants*,
https://doi.org/10.1007/978-981-15-6918-0_2

2.1 Potential Economic Impact of Innovation

In the preceding chapter, the field of biomaterials science is introduced. Many key examples are provided to demonstrate the critical use of implants in various disease treatments. For example, geriatric care is of major concern in India. The World Health Organization (WHO) estimated that 130 million people will suffer from osteoarthritis and 40 million people will be severely disabled by osteoarthritis by 2050. In India, 20% of the disabled population has a movement-related disability, accounting for about 53 lakhs, and this figure is increasing annually since 2011.

Currently, there are more than 300,000 cases every year of total knee replacements (TKR) and more than 100,000 cases every year of total hip arthroplasty (THA) procedures in India. The available articulating joint implants generally offer a trouble-free life for about 10–15 years, which is inadequate, considering the increased lifespan for humans in many developing nations. Therefore, a search for ideal prosthetic materials together with treatment methods, reconstructive solutions, and surface designs is currently being pursued in the field of orthopaedic biomaterials.

The increase in the number of revision surgeries is commensurate with the rate of increase in primary surgeries, and the majority of revisions are due to prosthesis failures—hence, the increasing demand for improving implants and thorough characterisation, globally. Trauma injuries due to accidents are another cause of mortality and disability. Although timely treatment is essential, the cost of surgery for cases of major trauma injuries or disabilities remains unaffordable to the majority of the population in many developing nations, due to unavailability of affordable implants, such as hip and knee implants, among other causes.

Significant socio-economic impact can be realised if we improve and develop the translational ecosystem that is the key to affordable healthcare in India.

2.1.1 International Market

According to industry estimates, the global biomedical devices market in 2020 is valued at around 470 billion USD, equivalent to 47,000 crores INR. The market is expected to grow at a CAGR of 7.5% over the next 5 years, as shown in Fig. 2.1. Globally, the principal markets are North America (50%), Europe (12%), Japan (10%), Germany (6%), and rest of the world (22%).

North America is the largest market for implants, and Asia–Pacific is the fastest growing market with India being a key player in the Asian market.

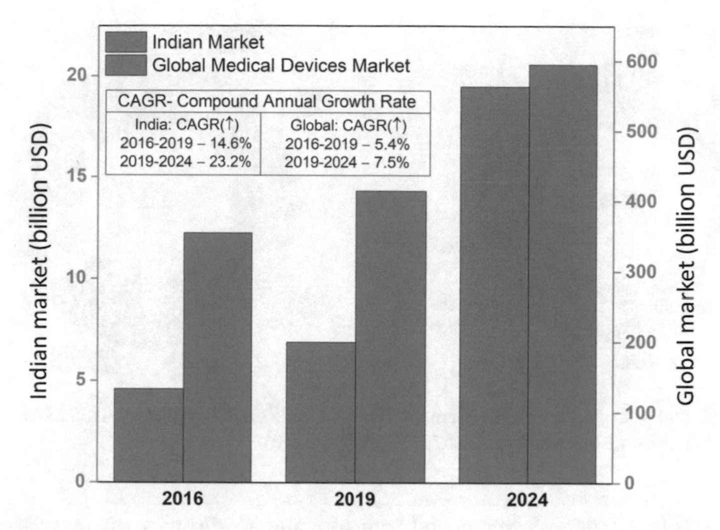

Fig. 2.1 Global medical devices market versus Indian medical devices market demand and forecast, 2016–2024 (*Source* Market Study of Advanced Ceramics and Composites in the Healthcare Space, Frost and Sullivan, 2020)

Out of the total healthcare expenditure in Europe, around 7.2% is attributed to medical technologies. In 2017, more than 13,000 patent applications were filed with the European Patent Office, in the field of medical technology. The European medical technology industry employs directly more than 675,000 people. There are almost 27,000 medical technology companies in Europe. Europe's top export locations for medical devices were USA (40.7%), China (9.6%), and Japan (6.6%).

The general trend of the bone implants market is shifting from autograft (autologous bone) to allograft (synthetic bone). In the case of the autograft, the acceptance by the body is very rapid, but the time involved and pain created is much greater than when compared to that of the allograft. Inferences on the implant market for treatment of specific human diseases can be realised only when one compares the market growth in that specific implant domain. The global orthopaedic market is approximately 30% of the total implants market and the market shares of segments in 2020, and as expected in 2023, are shown in Fig. 2.2. It is evident that relative changes in the segment shares for orthopaedic large joint implants, spinal implants, and orthopaedic trauma in 2023 are predicted to be marginal with respect to the present market status in 2020. A similar scenario is also expected for orthopaedic small bones and joints, orthopaedic biomaterials, and orthopaedic soft tissue regeneration. The overall trend therefore shows a steady, yet slow expected market growth in the THR application domain. The THR market is further subdivided into three sectors: total hip replacement, partial hip replacement, and revision and hip resurfacing. Overall, the total market segment share for orthopaedics is expected to remain around 70% and that for spinal, around 30% (see Fig. 2.2).

Top international manufacturers in the orthopaedic domain include Baxter, Becton Dickinson, Stryker, Cardinal Health, Abbot Lab, Siemens Healthineers, Royal

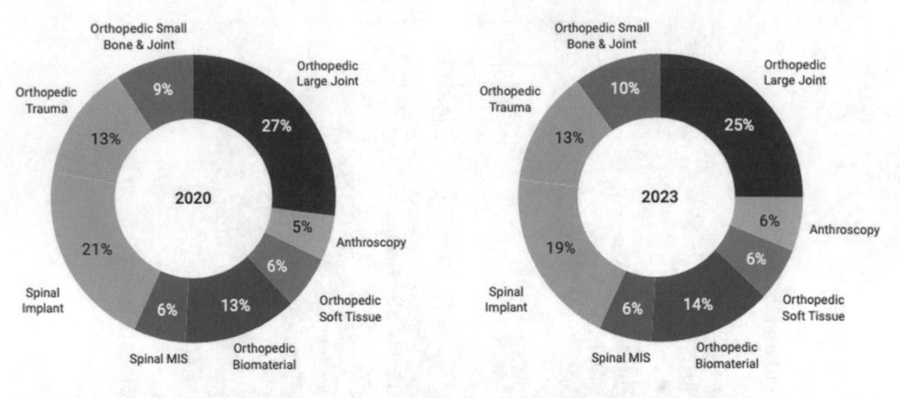

Fig. 2.2 Global orthopaedic market segment shares in 2020, and as predicted in 2023 (*Source* State of the Global Orthopedics Market 2019/20, iData Research)

Phillips, GE Healthcare, Johnson and Johnson, and Medtronics. As far as the manufacturing sector is concerned, the medical 3D printing global market is estimated to reach $1.2 billion by 2020 at a high compound annual growth rate (CAGR) of 29%. Medical applications for 3D printing include surgical models, bioprinted tissues, and medical implants, which account for 65% by value. Polymers are the preferred materials for 3D printing, and they account for about half of the market revenue in the materials market.

The Indian spine market does not have many domestic players. Fusion devices currently make up the largest market share for spinal devices, and the non-fusion device market share is expected to grow through 2019. Medtronic, Zimmer Biomet, and DePuy Synthes are dominating the market, while other players are Matrix Meditec, Inor Orthopaedics, SH Pitkar, and Biorad Medisys. Globally, the market for spinal implants is expected to grow at a CAGR of approximately 5.5% during 2017–2023, reaching more than $12 million USD during 2018–2024. The number of spinal procedures performed each year in the USA is over 1.4 million, with the greatest number being cervical fixation procedures (over 350,000 procedures yearly).

Interbody fusion procedures and thoracolumbar fixation procedures are the second and third most common in the musculoskeletal implant sector. The European spinal market size is approximately half that of the USA. Europe's fastest growing segment is the vertebral compression fracture market.

> The gradual rise in spinal disorders, including lumbar spine stenosis, degenerative discs, disc herniation, and spinal stenosis among others, is one of the major contributing factors for the growth of the spinal implant market.

The global spinal market, segmented by products, comprises spinal fusion implants, spine biologics, VCF treatment devices, spinal non-fusion implants, and spine bone growth electrical stimulation devices. The key players in the spine market are Medtronic, Johnson and Johnson, Stryker, NuVasive, Globus Medical, and Zimmer Biomet.

Dental implants are made up of the implant, the abutment, and the crown and can be used to replace single teeth or a set of teeth. They greatly improve the quality of life from the point of view of oral health. Available in different sizes and shapes, dental implants are employed according to the needs of the patient. The American Association of Oral and Maxillofacial surgeons has indicated that 69% of adults, aged between 35 and 44, could have lost at least one permanent tooth due to an accident, gum disease, failed root canal, or tooth decay. By age 74, 26% of adults have lost all of their permanent teeth. According to the market projections by GlobalData Medical Intelligence Centre, the market value for dental implants is expected to increase from $4 billion USD in 2020 to around $6 billion USD in 2028 (see Fig. 2.3). An important observation is that the market is growing steadily and the projection indicates the doubling of the market demand during 2018–2028 in the dental implant sector.

Looking at the larger global dental implants and prostheses market, it generated $7.2 billion USD revenue in 2017 and was predicted to increase at a CAGR of 6.1% during 2018–2024, reaching a value of about $12 billion USD in 2025 (Fig. 2.4). This, along with the rising geriatric population, provides an opportunity for key players to cater to the enormous untargeted population suffering from edentulism.

In the more advanced markets, computer-aided design/computer-aided milling (CAD/CAM) has revolutionised dental care and the patient experience. It allows the manufacture of crowns immediately within the clinic or hospital. Dental tourism is evolving in emerging economies, such as India. Increasing economic stability and dispensing power are further expected to accelerate growth of the market.

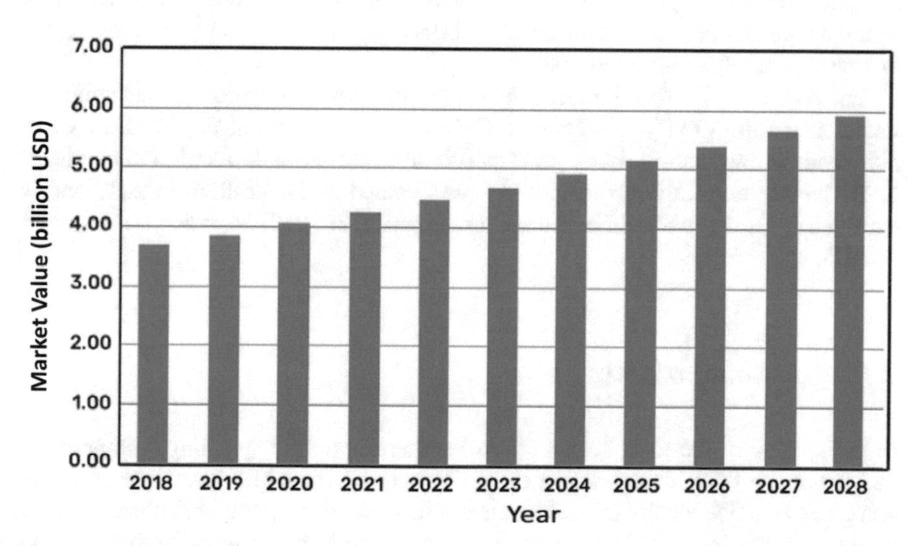

Fig. 2.3 Global market prediction: dental implants market (*Source* GlobalData Medical Intelligence Centre)

	Dental Implants and prostheses	Craniomaxillofacial devices and implants
Global Market Size (2017):	$7.2 billion USD	$1.4 billion USD
Expected Growth During 2018 – 2024:	CAGR of 6.1% to reach a value of around $12 billion USD by 2025	CAGR of 8.6% to reach a value of around $3 billion USD by 2025

CAGR: Compound Annual Growth Rate

Fig. 2.4 Market size for dental and craniomaxillofacial implants (*Source* GlobalData Medical Intelligence Centre)

With respect to end-users, the customer base is made up of dental clinics and hospitals, dental laboratories, dental colleges, and research establishments.

On the basis of dental implant type, the market is divided into tapered and parallel-walled dental implants, which are manufactured from titanium and lately, zirconia blanks.

The tapered implants sector share is 77.7%, and that of the parallel-walled implants is 4.6%. The zirconia implant segment is forecasted to show significant growth, which is expected due to the growing adoption of these implants (see Fig. 2.5). Major players in the dental market are Straumann AG, Dentsply Sirona, Zimmer Biomet, Henry Schein, and the Danaher Corporation.

The craniomaxillofacial (CMF) device product segment encompasses temporomandibular joint (TMJ) replacement, CMF distraction, cranial flap fixation, CMF plate and screw fixation, bone graft substitutes, and thoracic fixation. The global CMF devices and implants market size was valued at $1.4 billion in 2017 and is expected to grow at a CAGR of 8.6% over the period 2018–2024, reaching $3 billion in 2025.

2.1.2 *National Market*

In India, 30% of the total healthcare expenditure is public spending, compared to 48.3% for the USA. Just 1.4% of the GDP is spent on public healthcare in India, compared to 8.3% for the USA. India needs to increase its public healthcare expenditure. Considering that public healthcare is often the only recourse for the rural and poor population, the public healthcare expenditure is expected to reach ~3% of GDP by 2025.

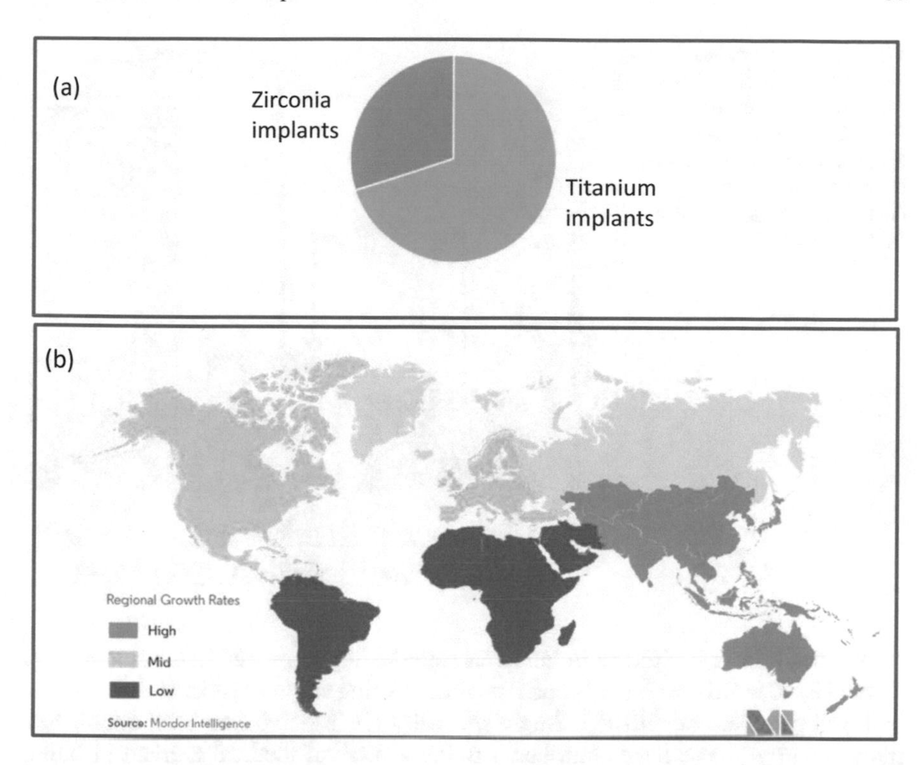

Fig. 2.5 Dental implants market sectors, revenue share (%) by type of implants (**a**) and growth rate at different geographical regions in the world (**b**) (*Source* Dental Implants Market—Growth Trends and Forecasts (2020–2025), Mordor Intelligence)

The business opportunity of medical devices in India is attracting not only multinational companies established in the area, but also corporate houses with presence in related areas, such as pharmaceuticals, industrial ceramics, and precision machining. The new entrants are exploring domestic acquisitions to gain a foothold in the market and also international acquisitions to gain product and technology advantage.

In India, the industrial sector needs to improve its R&D capabilities in order to collaborate effectively with academic institutions and national labs.

As in 2019, the medical device market in India was valued at around 15 billion USD (see Fig. 2.1). The evolving market structure calls for companies to adapt their strategies to participate in the growth opportunities, as the Indian market emerges as a 20 billion USD business in 2024 (see Fig. 2.6). India is one of the fastest growing medical devices markets globally—it has threefold higher CAGR compared to global growth (Figs. 2.1 and 2.6). The Indian market is the fourth largest market in Asia.

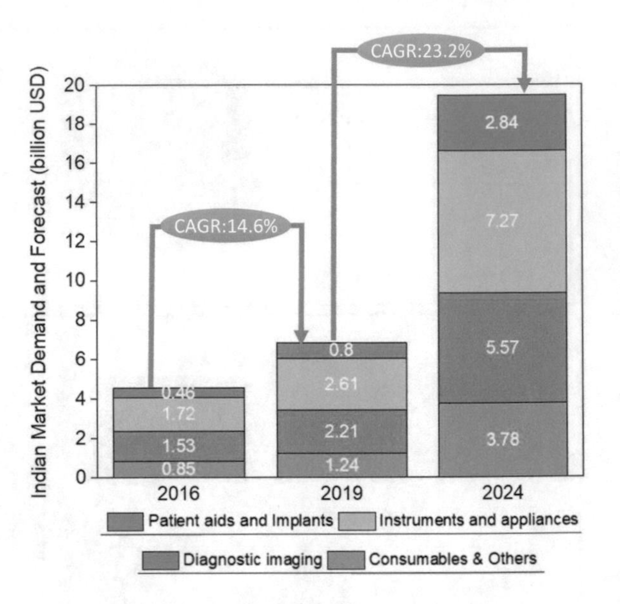

Fig. 2.6 Indian medical devices market demand and forecast—by segments, 2016–2024 (*Source* Market Study of Advanced Ceramics and Composites in the Healthcare Space, Frost and Sullivan, 2020)

The delivery of healthcare in India has significantly improved in the last decade. Several large multi-speciality hospital chains, offering world-class treatment options and facilities were established. Single speciality chains have also been established more recently. These have contributed to the growth of medical tourism in India, which is positioned to further grow, given the quality and cost-effective care in world-class accredited hospitals, such as Narayana, Hinduja, Fortis, and Apollo Groups. Imports are growing rapidly as world-class hospital groups build high-end infrastructure and open India to medical tourism, a growth which now adds 2 billion USD to the Indian healthcare market. The National Medical and Wellness Tourism Board has been recently constituted by the Ministry of Tourism in India, for promotion of medical tourism and for support of the medical/wellness segment.

Many international patients travel to India every year for complex surgeries, such as cancer treatments, while others travel for affordable procedures, such as plastic surgery that may not be covered under patients' insurance. India's affordable medical treatments, high-quality healthcare services, skilled medical staff, traditional medicine, wellness centres, and sightseeing attractions have made the country a preferred destination for medical tourism.

The global medical tourism industry was valued at $76 billion USD in 2018 and is expected to grow further at an average growth rate of 20%. Currently, India holds approximately 18% of the global medical tourism market (approximately $14 billion USD). Major medical tourism destinations are Thailand, Malaysia, Singapore, India, South Korea, Brazil, Costa Rica, Mexico, Taiwan, and Turkey. Medical tourists travel to these countries to avail of affordable and high-quality medical treatments. The percentage of cost savings by country with respect to similar treatment in the USA is shown in Fig. 2.7. Also, India is perceived as a preferred destination for joint and cardiac surgeries by several neighbouring Asian countries and many African nations.

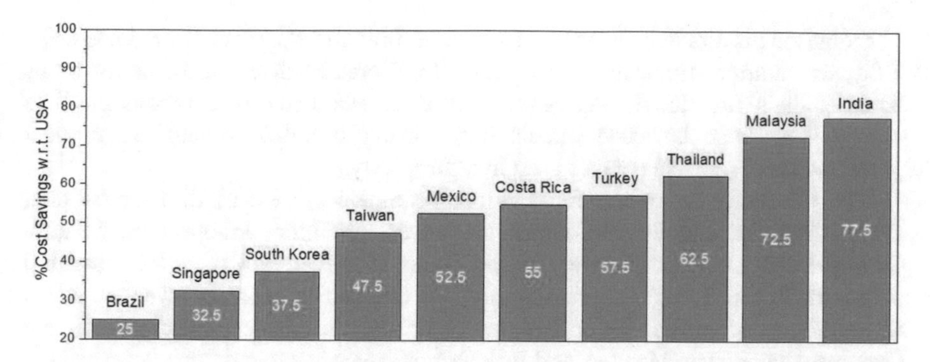

Fig. 2.7 Indian medical tourism market: percentage of cost savings from medical treatments by region compared to the USA, 2018 (*Sources* Ministry of Tourism India; Patients Beyond Borders; India Brand Equity Foundation; Market Study of Advanced Ceramics and Composites in the Healthcare Space, Frost and Sullivan 2020)

The medical tourists generally save around 77.5% of treatment costs in India, compared to what they incur if they undergo similar medical treatment in the USA.

Significant growth of the market (118%) is expected between 2020 and 2024. Approximately 214 million patients are suffering from orthopaedic and dental-related diseases. With respect to the need for dental treatment, 11–37% of edentulism occurs in the age group of 65–75 in India. Dental caries is occurring at a rate of 70–80% in the adult population.

The orthopaedic market is split evenly between the joints and trauma–spine segments. The joints market in India has a predominance of foreign MNCs, whereas the trauma–spine market is highly fragmented with a number of Indian companies occupying the majority of the market share (revenues greater than $8 million USD). Indian companies need CE marking and US FDA approvals to penetrate into global markets. There has been a recent trend with multinational companies bringing in Chinese products to compete with Indian players on price. Over 25,000 total joint replacement procedures were registered in 2016 at over 740 hospitals in 200 cities across India. The orthopaedic joints market is composed predominantly of hip and knee segments.

The knee sector is predominantly dominated by multinational players, such as Zimmer Biomet, DePuy Synthes, Stryker, and Smith and Nephew. Currently, their products are better suited for western populations.

In view of different patient anatomy in Asian populations, effective customisation of biomaterial implants is needed to meet the local requirements.

Domestic players include Meril Healthcare, Inor Orthopaedics, Biorad Medisys, TTK, and Sharma Orthopaedics. For example, Biorad Medisys manufactures Indus Knee, India’s first totally indigenous posterior stabilised knee prosthesis. The company has also launched India’s first registry initiative to map the clinical performance of medical devices used in arthroplasty.

The hip sector has major multinational companies, such as Zimmer Biomet, DePuy Synthes, and Stryker. Domestic players are Inor Orthopaedics, Sharma Orthopaedics, Meril Healthcare, and SH Pitkar. However, they often lack total hip solutions. In this market, approximately 200 companies generate revenues worth $3 million to $5 million USD. Notable companies in this sector include Sharma Orthopaedics, Matrix Meditec, and Inor Orthopaedics.

In the past decade, a few companies have set up innovation centres in India and manufacturing units for medical devices. US-based Orchid Orthopedics has opened a new manufacturing site in Pune and plans to serve the Asian and global markets by providing additional manufacturing options. Stryker is interested in developing the emerging markets of India and China and has set up an R&D centre in Gurugram to develop products more suitable for Asian markets. According to the company, emerging markets can contribute to as much as 12–14% of its global sales over the period of 2016–2021. Zimmer Biomet and Indo-UK Institutes of Health (IUIH) are planning to build the Biomet Institute of India, which will train more than 1000 orthopaedic surgeons a year on new procedures. DePuy Synthes established the DePuy Institute for Advanced Education and Research in Chennai, which is expected to train several hundreds of surgeons a year on the latest technologies.

In the national market, almost 80–85% of demand is met through imports. Imports are highest in patient aids, implants, and instruments and appliances, while it is lowest in consumables. Consumables and Diagnostic Imaging sectors dominate the device market, occupying more than 50% of the market. There are around 750–800 device medical device manufacturers, with around 65% of them operating in the consumables segment. The Indian orthopaedic and prosthetic device market is valued at $660 million USD and the dental product market at $290 million USD in 2020 (see Fig. 2.8). The overall projection is that the combined market will grow to 2.84 billion USD in 2024, as shown in Fig. 2.8.

In the coming phase of the Indian medical device industry, the products of domestic companies are expected to be on par with imported products, to gain market share despite the presence of foreign multinational companies, and also to emerge as strong contenders in the rest of world export market. For example, domestic players with products competing for key segments are Biorad Medisys’ knee/hip replacement systems and Maxx Medical’s Freedom Knee Replacement Systems.

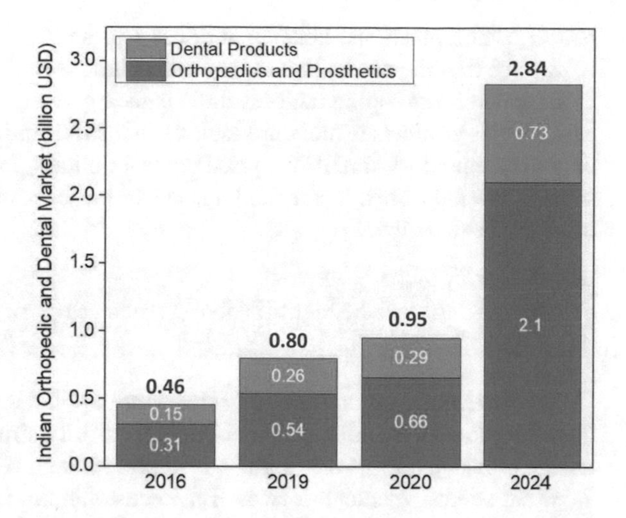

Fig. 2.8 Indian orthopaedics and prosthetics and dental products demand and forecast, 2016–2024 (*Source* Market Study of Advanced Ceramics and Composites in the Healthcare Space, Frost and Sullivan 2020)

2.2 India's Healthcare Initiatives and Translational Research Facilitation

Not only are the innovations important, but also, the initiatives which improve access to healthcare are vital. The Ayushman Bharat Scheme is a national health coverage plan for 500 million underprivileged citizens—through the scheme, the Indian government has delivered INR 79.01 billion worth of free secondary and tertiary treatment to beneficiaries across 224 hospitals at the national level (coverage up to 5 lakhs per family). Under this scheme, population coverage may increase from the current, approximately 40% of the nation, to as high as 75% in the next few years. The Union Ministry of Health and Family Welfare plans to launch 'My Health Record', a mobile app, which will support digital health and telehealth in India.

In 2015, the Government of India launched the initiative, Startup India. Subsequently, more than 3000 start-ups have been supported by seed funding from state governments, via grants to incubators, loan interest subsidy, interest free loan, reimbursement of subsidy on lease rental and funds as monthly sustenance allowance, product development assistance, and marketing assistance.

The Ayushman Bharat Pradhan Mantri Jan Arogya Yojana (AB-PMJAY) Start-Up Grand Challenge is a national initiative, that intends to stimulate the Indian start-up community to generate solutions for empowering people to gain access to affordable healthcare.

Launched by the National Health Authority in partnership with BIRAC and Startup India, start-ups engaged in areas such as medical devices, digital health,

health communications, hospital services and hospital management, and medical workforce training are invited to innovate solutions in the healthcare sector.

Like other developing nations, India is facing several market-related challenges. The market limiters in India are lack of industrial initiatives, such as stakeholder network, regulatory initiatives, price control policies, lack of appropriate business models, low public healthcare funding, and the presence of competitive global players in biomedical devices.

The implants, that are highly expensive, run the risk of poor market acceptance.

The government is considering price control policies as an option to bring down costs for patients. In order to have control on the price structure of imported implants, recently, the Government of India has implemented a rise in import duty from 5 to 7.5% on several medical devices. Basic customs duty was lowered to 2.5% and there has been an exemption from special additional duty (SAD) on raw materials, parts, and accessories for the manufacture of medical devices. However, there are mixed opinions as to whether this will help patients in the long term as this might disincentivise both domestic and multinational companies from pursuing innovation. A challenge for smaller players is not being able to build an influential network among physicians and the advantage is for larger players, who are leveraging their network with the medical fraternity and academia.

In Europe, medical devices are classified into three major classes (based on risk and fault in functioning), with Class I being the lowest risk and Class III being the highest risk. The Indian regulatory system should take safety as a major priority and work towards such a classification, so that appropriate measures should be taken as per the risk level. Also, health technology assessment (HTA) by the responsible agency of each country is playing an important role in determining the appropriate usage, coverage, or reimbursement for medical technologies at both national and regional levels in Europe.

2.2.1 Translational Research Institutes and Programmes

India is transitioning into a nation with government-funded research parks. Under the 'Make in India' initiative, the Government of India has been actively considering the establishment of four medical device parks to be set up in Andhra Pradesh, Telangana, Tamil Nadu, and Kerala. Further details on this initiative are shown in Fig. 2.9. These parks will provide the necessary infrastructure for research, while facilitating easy access to standard testing facilities, and manufacturing of high quality, yet affordable medical devices. The facilities are expected to include component testing centres, biomaterial/biocompatibility testing centres, engineering design, and 3D printing

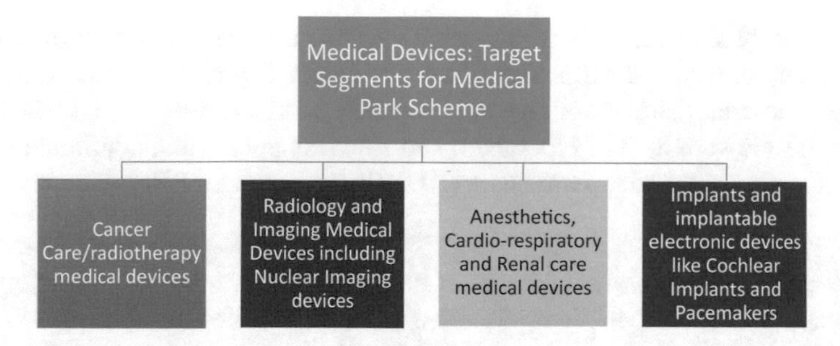

Fig. 2.9 Key device and implant segments that are being targeted by the Government of India's Medical Park Scheme (*Source* www.MansukhMandaviya.in)

facilities for medical grade products, sterilisation, and toxicological testing centres. The key device and implant segments that are being targeted are shown in Fig. 2.9.

The Indo-UK Institute of Health (IUIH) had discussed the idea of 11 med-cities with the Indian government in 2015. The med-cities were intended to be built in Punjab, Gujarat, Andhra Pradesh, Rajasthan, Karnataka, Uttar Pradesh, West Bengal, Maharashtra, Madhya Pradesh, Haryana, and Telangana. The institute is partnering with medical technology companies, which will be carrying out manufacturing at the med-cities. Each med-city was intended to contain a centre of excellence in a single area of expertise. However, as previously discussed, it appears that the government is more interested in comprehensive and self-sufficient facilities, rather than those specialising in particular areas. An overview of the 'Make in India' Initiative is shown in Fig. 2.10.

Fig. 2.10 Make in India Initiative—for supporting the Indian manufacturing ecosystem (*Source* Market Study of Advanced Ceramics and Composites in the Healthcare Space, Frost and Sullivan, 2020)

Of the established research initiatives, IKP Trust is a major contributor to India's research output. Under the Startup India Initiative for Higher Education Institutions (SIIHEI), the establishment of five new research parks is underway (IIT Delhi, IIT Kanpur, IIT Guwahati, IIT Hyderabad, and IISc Bangalore) and continued funding for two already approved research parks (IIT Kharagpur and IIT Bombay) is ongoing.

On the research facilitation frontier, Kalam Institute of Health Technology and the Andhra Pradesh MedTech Zone are notable examples in India.

The participation of academia is indeed important in all such national initiatives. The Healthcare Technology Innovation Centre (HTIC) is an example of an existing centre established by academia and the government. Located at the Indian Institute of Technology Madras Research Park, Chennai, HTIC is an initiative of Indian Institute of Technology Madras (IITM) and Department of Biotechnology (DBT), Government of India. HTIC collaborates with leading medical institutions and a range of industry players and facilitates the collaboration of technologists, engineers, doctors and healthcare professionals, industry and the government.

The Translational Health Science and Technology Institute, THSTI, is an important example of a translational research ecosystem in a national institute in India.

Established in 2009, THSTI undertakes research in biotechnology solutions for highly prevalent human diseases in India in the areas of infection and immunology, maternal and child health, non-communicable diseases, and multidisciplinary clinical and translational research.

Several clinicians are involved in research at THSTI. The objective of connecting academia with THSTI is to combine engineering sciences with biological sciences. This would facilitate the framing of clinically relevant strategies to develop materials for implant and drug delivery applications, followed by clinical trials and commercialisation.

2.2.2 Kalam Institute of Health Technology and the Andhra Pradesh MedTech Zone

Kalam Institute of Health Technology (KIHT), Visakhapatnam, was set up as a society under the Department of Biotechnology, Government of India, in 2017 (https://kiht.in/). KIHT's mandate is to facilitate focused research on critical components pertaining to medical devices by supporting stakeholder institutions engaged in development and promotion of healthcare technologies. KIHT has a unique e-auction platform for accumulating healthcare technologies across the country and creating a repository technology bank and facilitating transfer of such technologies/prototypes/ intellectual property to interested manufacturers and incubators.

AMTZ is a 270 acre-zone for medical device manufacturing. KIHT is located in the AMTZ campus.

The AMTZ group of institutions is the only comprehensible model in India providing innovation support, testing, validation, and manufacturing infrastructure along with policy support in health technology sector, within one ecosystem.

After its inception, AMTZ held several domestic and international investor meets to promote domestic manufacturing and awareness about the immense growth opportunity that the medical device sector offers.

KIHT provides research and development support, market intelligence and trade support as well as market access support. AMTZ's Centre for MedTech Innovation and Rapid Prototyping is designed to provide further support to researchers for development of prototypes, including customisation and design. It contains a 3D printing facility, to cater to start-ups, medium-scale and small-scale enterprises as well as manufacturing sectors. The advantages of the 3D facility are mass customisation, supply chain simplification and waste reduction.

KIHT/AMTZ's facilities include the centre for gamma irradiation, biomaterials testing facility, pre-clinical facility and animal house, electromagnetic compatibility (EMC) testing, electromagnetic interference (EMI) testing, and acoustic facility.

The gamma irradiation facility set-up at AMTZ with cobalt source supply from BRIT/BARC would cater 15% (INR 4500 crores) of medical device market. A multipurpose plant was designed as per the guidelines of the Atomic Energy Regulatory Board (AERB).

AMTZ is also a site for incubation of start-up companies. KIHT is the official policy body for all departments of government for all health technology avenues. MediValley, funded by NITI Aayog, the Planning Commission for the Government of India and BioValley, funded by the Ministry of Science and Technology are world-class incubators, representing the largest incubator ecosystem in the country in the realm of healthcare technology.

MediValley handholds innovators along all stages, from ideation to manufacturing while allowing full access to the AMTZ group of institutions.

The Fourth Global Forum on Medical Devices, organised at AMTZ in December 2018, was dedicated to discuss standards, regulatory framework, and medical access. In order to bring together mentors, financiers, distributors, insurers, and all stakeholders together for medtech start-ups, AMTZ hosts a quarterly start-up day in collaboration with the Confederation of Indian Industry (CII). AMTZ held a joint workshop on 3D bioprinting with think3D in May 2019, which featured speakers from reputed Australian and Indian institutions.

AMTZ was cited at the World Economic Forum 2019, Switzerland, as being a sector-specific innovation cluster of global excellence. Such clusters enable governments to plan training programmes in collaboration with local industry and academia. In April 2019, AMTZ hosted a workshop on medical devices and diagnostics for the SEARN. In May 2019, a workshop on medical device regulation and certification in the USA and Europe was held, thereby enhancing the knowledge of these prominent regulatory systems in India. KIHT-AMTZ have 4 international MOUs, and 93 memberships in various national and international bodies.

The Make in India leadership institute (MIILI) programme, offered by KIHT, is a rigorous executive programme of three weeks duration for working professionals and exceptional graduates in the medical technology field. The course covers knowledge in several thematic areas in medical devices: regulation, financing, pricing, market access, R&D, and standardisation. The course broadly offers content of seven modules, apart from activities like a medical technology industrial tour, webinars by renowned international faculty members and a leadership session by an eminent industry champion.

KIHT-AMTZ launched a novel Patent Academy, which offers a wide range of Intellectual Property Rights (IPR) courses designed exclusively for students, IP career seekers, researchers, entrepreneurs, and industry segments. In another initiative, biomedical practitioners are encouraged to take the Indian Biomedical Skill Certificate (IBSC) test, established by AMTZ and supported by the Quality Council of India (QCI) and the Association of Indian Medical Devices Industry (AiMeD).

2.2.3 *NITI Aayog and the National Health Stack*

The Ayushman Bharat Yojana is a national programme designed to establish 1.5 lakhs of health and wellness centres for comprehensive primary healthcare offering preventative and promotive healthcare accessible to all.

The second arm of the Ayushman Bharat programme is the Pradhan Mantri-Rashtriya Swasthya Suraksha Mission (PM-RSSM), which will cover more than 10 crores of poor and vulnerable families providing coverage of up to 5 lakhs per family per year for secondary and tertiary care hospitalisation. The design of the Ayushman Bharat Yojana was released in the Union Budget of the fiscal year of 2018–2019 and is expected to be mediated by the states of India.

The National Health Stack (NHS) is conceived as a digital infrastructure for implementing health initiatives, such as managing health data, managing coverage and claims, and enabling patients to access their own health data and national health analytics.

The platform could help to improve process flows for service providers. Other potential building blocks include policy repositories, national level registries, workflows, incentivised access to stakeholders through open application program interfaces (APIs) and consolidation of all insurance and assurance claims for healthcare at all levels (primary, preventive, secondary, and tertiary). The NHS is expected to ensure deployment of patient-centric policies.

The current Indian healthcare system is lacking in standard treatment guidelines. There is a need to create consistent and routine clinical processes with accountability structures and measurable improvements in quality of care, focusing on patient safety, comfort, satisfaction, and clinical outcomes. Therefore, the data systems should ensure monitoring and verification of processes and patient outcomes.

Dexterity, flexibility, and evidence-based smart policy making are required for the implementation of healthcare initiatives, such as the NHS in India, which is a digital framework usable by centre and state across public and private sectors. The initiative of the NHS is expected to be a technology-informed model for universal health coverage, enabling cost-effective healthcare for all. The programme aims for digital health records for all citizens by 2022.

Through the NHS, NITI Aayog is intending to shape strategic policy actions to improve health security by operating in mission mode, what would be the largest healthcare programme in the world.

The NHS would draw from big data analytics, machine learning, and artificial intelligence. Therefore, NHS in India is perceived to be the best-in-class healthcare information technology system.

2.3 Education and Human Resources

Education and human resources are key to the success of any translational research initiative. The Indian medical devices industry has not operated like those of the established economies of developed countries. In the developed country model, the companies employ the most expensive talent, such as the best engineers and researchers, and then recover the money via end product pricing and selling to insurance companies. In India, the talent available is variable in quality and the recipients of the service are small Indian companies, who often cannot afford to pay a high salary to many of their manpower.

> With the anticipated growth of the Indian market, there is a distinct need to revamp the bioengineering and medical curricula in academic institutions.

With a view of attracting bright and meritorious students to research, the Prime Minister's Research Fellows (PMRF) Scheme has been launched recently. The scheme seeks to admit meritorious students into PhD programmes at IITs, IISc, and IISERs. These scholars receive fellowships, which are comparable or better than those of graduate students in the USA or other western countries. More than 200 fellows have been admitted to the scheme so far.

2.3.1 Bioengineering Curriculum and Indian Academia

At the backdrop of the accelerated growth of the field of biomedical engineering in the western world, this interdisciplinary field is growing both in the context of academic curriculum and translational research.

> Many universities around the world have already started new interdisciplinary academic programmes namely, under the titles on 'biological engineering' or 'bioengineering' or 'biosystems science and engineering' or 'medical science and technology' or 'clinical engineering'.

Such academic programmes will provide a comprehensive platform for researchers from multiple disciplines to interact while collaborating to share complimentary expertise. On the brighter side, this practice is followed in some IITs and a few schools in India. However, currently in India, there is a dearth of widespread initiatives in academia like study programmes and training of young researchers in the area of biomaterial sciences. Hence, many students lack the knowledge about the fundamental concepts of biomaterial sciences, which can be a bottleneck to pursue research in this area.

It has been widely perceived nationally that undergraduate/postgraduate studies in biomedical engineering (BME) or similar disciplines still do not attract the best students. This can be largely attributed to, (a) poor job market in India, (b) lack of fundamental concepts in undergraduate courses as well as common curriculum of BME and (c) lack of integration among the stakeholders, national laboratories, hospitals, industries, and universities/IITs/IISc. The current employment for many undergraduate students of BME is seen in non-technical areas, like customer support, sales and services, maintenance, and management of imported healthcare equipment. Also, many undergraduate students of BME opt for higher studies in the field of bioengineering, abroad.

Considering the increasing importance of high-quality manpower in the healthcare sector, a new Graduate Aptitude Test in Engineering (GATE) paper in biomedical engineering was introduced and has been effective from 2020 for screening of undergraduate students before they can be admitted to postgraduate programmes in India. It is envisaged that this initiative will enable the academic institutes to modify the course curriculum to introduce a problem-solving/analytical approach at the intersection of biology, medicine, and engineering. Also, more focus is expected to improve skills in the areas of anatomy, physiology, molecular biology, and human–machine interface, etc.

2.3.2 Medical Education and Research in India

Progress in the field of biomaterials science and implants can be accomplished only with commensurate progress in medical education and the clinical research ecosystem.

> In 2017, there were 479 medical colleges in India, with admission capacity of over 60,000 at the undergraduate level.

The Medical Council of India (MCI) is the government-mandated regulatory agency for medical education. The total admission capacity for postgraduate medical degrees is about 8684 in MS, 16,195 in MD, 1280 in MCh, 1420 in DM, and 3837 in various diploma disciplines.

The doctor—population ratio varies from state to state with unequal distribution between rural and urban areas. The challenge of promoting a spirit of dedication for serving rural areas among medical graduates is still being faced. The World Health Organization's South-East Asia Regional Office in conjunction with Reorientation of Medical Education (ROME) suggested that to solve this problem, the medical education community, including residents, interns, and students should be oriented to the conditions existing in rural communities and that training in management of health problems in these areas should be provided. The scheme aimed to render comprehensive healthcare in villages in collaboration with the concerned primary healthcare centres. Each medical college was to establish a well-knit rural referral system, engage in outreach activities, post medical students in the community, provide mobile clinics, and stimulate the entire faculty to conduct community-based training.

In order to improve the distribution of research capabilities across the nation, ICMR has given substantial funds to 100 medical colleges in India towards research and have also started funding small ICMR research units across the country with instrument and manpower support. Also, research funding organisations are spending 10% of their budget to develop research infrastructures in North-Eastern states of India. The medical universities or national institutes of importance, like AIIMS can significantly contribute to research on biomaterials and implants, by providing the ecosystem for better interaction with researchers from academia and for conducting pre-clinical and clinical studies. The academic researchers would be largely benefited from regular interactions with clinicians or clinical scientists and would gain insights into the clinical relevance of a given research problem.

Although conducting randomised controlled trials (RCTs) would be advantageous financially, research capability and reputation of institutions, rules and regulations associated with conducting an RCT are very restrictive.

Another important issue that is raised is the paucity of primary healthcare physicians, coupled with the excess of specialists—an imbalance that needs to be corrected. The MCI has recently developed guidelines and regulations pertaining to the establishment of new medical colleges, the opening of higher cases of study, and a greater control of admission to medical colleges. Nevertheless, with the sudden increase in private medical colleges, there has been a concern regarding the quality of education and also issues with maintaining a part of the teaching approaches relevant to rural community constraints. Also, it is a current challenge that medical students should be more able to respond to local health problems and the related social, cultural, and economic factors, by meaningfully applying knowledge and analysing facts.

Regarding quality of teaching, there are inadequate opportunities for teacher training and orientation, and lack of incentive structures to recognise and reward teaching effort. A group of medical schools in the country led by four leading institutes has organised a Consortium of 20 medical colleges for undertaking innovative projects in medical education. The Consortium has established a collaboration with the Department of Medical Education, University of Illinois, Chicago, USA.

The leading institutes in clinical research are All India Institute of Medical Sciences (AIIMS), New Delhi; Institute of Medicine, Banaras Hindu University

(BHU), Varanasi; Christian Medical College, Vellore; and Jawaharlal Institute of Postgraduate Medical Education (JIPMER), Pondicherry and, Postgraduate Institute of Medical Education & Research (PGIMER), Chandigarh.

The Consortium plans to conduct advanced courses for a Fellowship in Medical Education Scheme, in order to develop educators that will lead educational changes in India. The MCI has recognised 10 nodal centres for this purpose. Also, the points raised at the Edinburgh Declaration of the World Federation for Medical Education have been fully endorsed at the national level. There has been a suggestion for the planned constitution of a Medical Education Commission, and also the MCI has recently attempted to develop a needs-based curriculum. Revision of the rules and regulations governing undergraduate and postgraduate education are underway by the MCI. In future, it is expected that the National Medical Commission will guide medical education and healthcare in India.

There exists an unmet need to revamp the medical and bioengineering curricula and research ecosystem in India, and a similar situation prevails in many developing nations.

In this context, the importance of key roles in building research teams has been highlighted by the Stanford University Biodesign Leadership Programme in their review of the first twelve years of the programme. Four personality profiles of highly functional teams were highlighted, and these are the builder, who is responsible for design and prototyping, the organiser, who monitors progress, the researcher, who will distil information from the body of clinical, engineering and business literature and the clinician, who understands the complex issues associated with introducing a new technology into clinical practice. It is hoped that such schemes will be considered, while introducing new medical education and research policies in India.

2.4 Funding Status and Opportunity

2.4.1 National Funding Schemes

This section highlights the status of funding opportunities for basic research and translational programmes in the area of biomaterials science and implants funded by the Government of India.

2.4.1.1 Ministry of Science and Technology Funding

In India, the research on biomaterials science, biomedical engineering, or bioengineering-related areas is significantly funded by the federal agencies, such as the Department of Science and Technology (DST), Science and Engineering Research Board (SERB), and Department of Biotechnology (DBT), and the Indian Council of Medical Research (ICMR). In addition, CSIR, DRDO and DAE also fund such research activities mostly to the scientists, who are working at the national laboratories affiliated with the respective organisations, and the academic researchers, in the form of extramural grants to a much less extent (quantum of funds for bioengineering). DRDO has a Life Science Research Board (LSRB), which administers funding related to bioengineering areas.

Two different panels of SERB, i.e. Materials, Mining & Minerals Engineering; and Biomedical and Health Sciences panel, regularly invite proposals, fund a few selected projects, and also administer their progress. According to data received from the DST-materials panel, a total of 28 projects, worth Rs. 12.8 crores (≈1.8 million USD) were funded in the timespan of 2014–2019. These projects are largely on biomaterials development with a focus on advanced manufacturing, microstructure-property correlation and biocompatibility of Mg-based biodegradable alloys, Ti-based alloys (e.g. Ti6Al4V), bioceramics/bioglasses, and biopolymers/natural silk-based scaffolds. The outcomes of the funded projects were able to address some of the existing clinical challenges in orthopaedics, dentistry, cartilage regeneration, and neurosurgical treatment.

A similar review of funding from DBT revealed an impressive total of Rs. 40.8 crores (≈5.9 million USD) for research projects related to biodesign, bioengineering, or biomedical engineering-related areas over 2014–2019 (Fig. 2.11). Among these projects, several projects, particularly related to biomaterials and related devices, were worth of 23.5 crores (~3.4 million USD). All these projects are aligned with DBT's mandate to facilitate, a) medical technology (medtech) innovation in India, b) entrepreneurship and research leadership in biomedical engineering, c) medtech products across larger cross-section of Indians with broader affordability, and d) translational research involving multidisciplinary teams (clinicians, researchers from academia/national laboratories and industry).

2.4.1.2 Biotechnology-Focused Funding

The biotechnology sector is recognised as one of the key drivers for contributing to India's economy target of 5 trillion USD by 2024. The Indian biotechnology sector is poised to grow exponentially over the next decade. India is among the top 12 destinations for biotechnology in the world, valued at 51 billion USD during 2018–19.

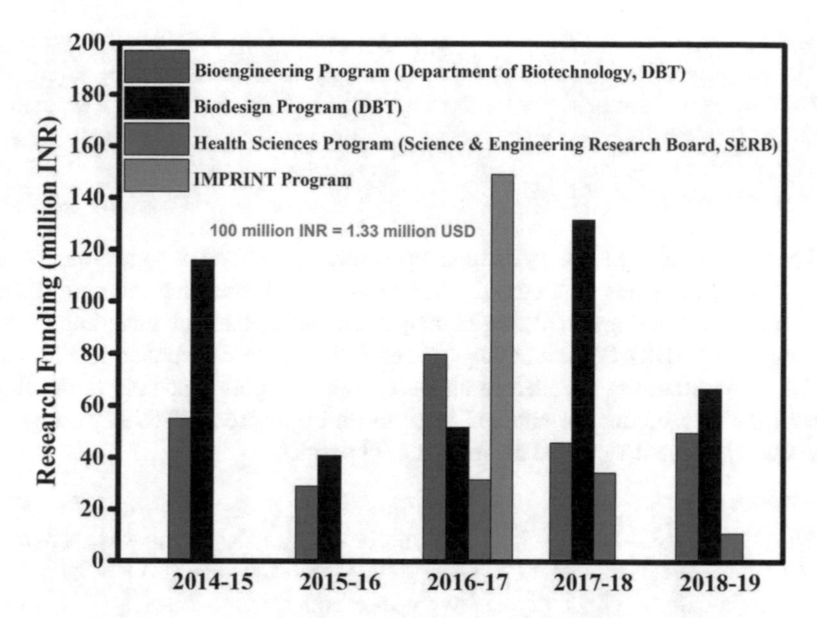

Fig. 2.11 Funding of research programmes on biomaterials and implants by the Government of India under several schemes monitored by DST and DBT

> The major policy initiatives of the Government of India (GoI), such as the Make in India programme are aimed to develop India as a world-class biotechnology innovation and biomanufacturing hub.

The Biotechnology Industry Research Assistance Council (BIRAC) is a not-for-profit public sector enterprise, set up by DBT, to strengthen and empower the emerging biotech enterprise to undertake research and innovation to address product development needs of national importance (https://www.birac.nic.in/). BIRAC implements a wide range of initiatives, such as providing access to risk capital, targeted funding, technology transfer, IP management, and handholding schemes that help bring innovation excellence to biotech firms.

There are several schemes of BIRAC/DBT, intended for different sectors. BIRAC's BIG scheme is intended for researchers/innovators in the area of biotechnology, who want to be entrepreneurs. The aim of this scheme is to provide support for proof-of-concept research. BIRAC has collaborated with the Society for Research and Initiatives for Sustainable Technologies and Institutions, Ahmedabad (SRISTI) for the BIRAC-SRISTI GYTI awards, which are intended to nurture 15 entrepreneurial ideas every year for two years.

The Small Business Innovation Research Initiative (SBIRI) scheme of the DBT is an enabling platform for the target organisations to realise their potential in terms of product and process development and taking products and technologies to the market.

The Biotechnology Industry Partnership Programme (BIPP) is a government partnership with industries on a cost-sharing basis for path-breaking research in frontier futuristic technology areas, having major economic potential, including Devices & Diagnostics. BIRAC's Promoting Academic Research Conversion to Enterprise (PACE) programme is intended to encourage and support academia to develop a technology or a product of national importance up to proof-of-concept stage and subsequently have it validated by an industrial partner.

The BIPP programme has the Academic Innovation Research (AIR) scheme to promote the development of proof of concept for a process or product by academia with or without the involvement of industry.

Alternatively, the Contract Research Scheme (CRS) aims at validation of a process or prototype (developed by academia) by an industrial partner.

Other than funding, BIRAC also provides support to innovators via the SPARSH programme (Social Innovation programme for Products: Affordable & Relevant to Societal Health). This scheme aims to create and foster a pool of social innovators ('SIIP Fellows') in the field of biotechnology and to provide a platform to share the best practices, understand the complex intricacies of business models in social innovation and network. BIRAC's Sustainable Entrepreneurship and Enterprise Development Fund ('SEED Fund') assists start-ups by providing capital assistance to start-ups with novel and worthy ideas, innovations and technologies. This would enable some of these start-ups to advance to a level, where they will be able to raise investments from Angel/Venture capitalists or to reach a position to seek loans from commercial banks/financial institutions.

The proposed seed support is intended to act as a bridge between promoters' investment and Venture/Angel investment. Also, available is the Accelerating Entrepreneurs (AcE) fund, intended to support start-ups.

BIRAC together with the National Biopharma Mission has initiated a technology transfer and commercialisation training programme, intended to strengthen overall capacity in the country in technology transfer.

In the last few years, a number of funding initiatives, in parallel to federal funding schemes, have come into the forefront to facilitate translational research in the biomedical engineering domain. One of them, is the Innovation, Knowledge, Progress (IKP) Trust, IKP. IKP is one of the six implementing partners of the Biotechnology Ignition Grant, supported by BIRAC (https://www.ikpknowledgepark.com/ikp-group.html).

The IKP trust has funded 300 start-ups and innovators, out of which 180 have been grant- or seed- funded, and 160 have been incubated, with 16 mergers and acquisitions being completed.

More than 7000 jobs have been directly created and 25,000 jobs have been indirectly created through all these entrepreneurial activities. IKP Trust has three programmes in its national footprint—IKP Knowledge Park, IKP Ventures, and IKP EDEN.

Within the IKP framework, a number of hubs or research parks are being established. The Science Park, Hyderabad, has a 200 acre campus with a 50,000 $ft.^2$ SME hub and 10,000 $ft.^2$ Life Science Incubator. The Life Science Incubator has been host to over 65 start-ups. It provides modular laboratory space, common equipment, funding and mentorship, legal and IPR support and facilitates networking among different incubators. In Karnataka, BioNEST is a MedTech Incubator in Bengaluru, and K-tech innovation hubs have six prototyping centres. IKP Knowledge Park manages several pan-India grants, with a total of 130 crores INR of funding. IKP Ventures manages an Indian innovation fund, which is currently dedicated to healthcare. IKP Engineering, Design and Entrepreneurship Network (EDEN) has a 25,000 $ft.^2$ hardware incubator and makerspace in Bengaluru for all streams of industrial design, which currently hosts over 80 start-ups. IKP EDEN has prototyping facilities for several applications, including medical devices.

IKP Knowledge Park's Grand Challenges Exploration (GCE) India Programme, in collaboration with BIRAC and the Bill and Melinda Gates Foundation, is designed to inspire out-of-the-box innovative ideas on high-priority global health issues, with funding of $100,000 USD for 18 months per project. IKP has an SME Hub at Hyderabad, operational from late 2019, mandated to provide 50,000 $ft.^2$ of wet laboratory space, 31 laboratory modules, and shared instrumentation facilities. In 2018, the GCE-India programme funded seven such programmes, two of which are in the areas of urological implants to treat bladder carcinoma and paper-based micro-fluidics for diagnostic purpose.

2.4.1.3 Recent Funding Schemes on Societally Relevant Areas

The Government of India has felt the persistent need to make the research in India more focused on the needs of society. Several schemes have been launched, such as *Unnat Bharat Abhiyan, Rashtriya Avishkar Abhiyan, Uchchatar Avishkar Yojna,* Pandit Madan Mohan Malaviya National Mission on Teachers and Teaching,

Swayam, and Global Initiative for Academic Network. The Science and Engineering Board, SERB, has launched 'SERB-SUPRA (Scientific and Useful Profound Research Advancement)' for scientists working in realms of highly unconventional pathways of transformative research, which may possibly lead to new hypotheses or challenge existing ones and provide innovative solutions.

Among various national initiatives, those of great importance discussed here are *Uchchatar Avishkar Yojna (UAY),* IMPacting Research Innovation and Technology (IMPRINT), and Scheme for Promotion of Academic and Research Collaboration (SPARC). UAY was launched in October, 2015 with an aim of promoting innovation of a higher order that directly impacts the needs of industry and thereby improves the competitive edge of Indian manufacturing. In UAY-I and –II, 84 and 64 projects were sanctioned for Rs. 254 crores (36 million USD/27 million GBP) and 135 crores (19 million USD/14 million GBP), respectively.

The Impacting Research Innovation and Technology scheme (IMPRINT) was launched by the Government of India in 2015 to provide solutions to the most relevant engineering challenges and to facilitate translating knowledge into viable technologies in 10 selected domains, including Advanced Materials and Healthcare.

An initiative of the Ministry for Human Resource Development (MHRD) and the Department of Science and Technology (DST) (https://imprint-india.org/), the aim of IMPRINT is to bridge the gap between the scientific knowledge base created through fundamental/applied research and enable translation of the research through engineering invention and technological innovation for national benefit. The domains that are relevant to biomaterials and bioengineering are healthcare technology, advanced materials, and nanotechnology. Notably, for this scheme, financial support from industry is necessary. The average funding for approved projects is around Rs. 2 crores per project (277,490 USD/227,290 GBP), and in 2018, 123 projects were sanctioned. In IMPRINT-I and –II, 142 and 126 projects were sanctioned for Rs. 313 crores (33 million USD/12 million GBP) and 112 crores (16 million USD/12 million GBP), respectively.

The Scheme for Promotion of Academic and Research Collaboration (SPARC) is an initiative of the Indian government, started in 2018 (https://sparc.iitkgp.ac.in/index.php). This initiative was based on the thought that the maximum benefit of a collaboration can be harvested only when the Indian research group, particularly the young researchers, can be trained at world-class research facilities in top international research groups. It funds long-term visits by the international faculty (2 to 8 months) and also funds the travel and sustenance of Indian students at one of the top 500 globally reputed universities/institutes. The programme aims to increase the number of highly trained scientific manpower in the country, and to promote Indian science and scientists in the world. The thrust areas in SPARC that are relevant to biomaterials and bioengineering, are affordable healthcare (convergence), nanotechnology, biotechnology and applications (innovation-driven), and design innovation (innovation-driven). Funding offered per project is in the following categories: up to Rs 50 lakhs, from Rs 50 lakhs up to Rs 75 lakhs, and from Rs 75 lakhs up to Rs 100 lakhs (100 lakhs is equivalent to 138,730 USD/113,639 GBP).

2.4.2 International Status

It would be interesting to briefly assess the funding status in the USA, UK, and other developed nations. The Engineering and Physical Sciences Research Council (EPSRC) is the main funding body for engineering and physical sciences research in the UK (https://epsrc.ukri.org/). The Biomaterials and Tissue Engineering Research Area constitutes 1.1% of the total EPSRC portfolio. The funding allocated for the area is £65.5 million (approx. 576 crores INR) for a total of 95 grants. The European Union's Horizon 2020 (2014–2020) research and innovation programme is another scheme, which funds projects in the area of biomaterials and bioengineering. Under the work programme 2018–2020—Future and Emerging Technologies (EUR 183.5 million, i.e. INR 1,524 crores for 2018–2020), the relevant area is 'Disruptive technologies to Revolutionize Healthcare'. Another relevant area, is 'Advanced Materials' under 'Key Enabling Technologies'.

The European Research Council (ERC), which funds research in the EU affiliated countries, has recently started a new initiative to fund 'frontier research' programmes (https://erc.europa.eu/). In general, the funding schemes of the ERC are tailored to different stages of the investigator's career, such as ERC starting grants, for early career researchers with 2–7 years' post-PhD experience, ERC consolidator grants for researchers with 7–12 years' post-PhD experience, and ERC advanced grants for established research leaders with a proven track record of research achievements. The maximum values of these grants range from €1.5 to 2.5 million for up to 5 years, the Indian equivalent being 12.42–20.7 crores INR. One can perceive that by categorising the levels of experience and achievements of the investigators, the ERC is enabling more of a level playing field among researchers over a wide spectrum of experience in science. Other grant schemes include the ERC Proof of Concept and the ERC Synergy Grants, of which the latter is for a team of up to four PI's groups.

The USA's National Institutes of Health (NIH) is the largest public funder of biomedical research in the world, investing more than $32 billion USD a year.

The National Science Foundation with an annual budget of about $7.8 billion USD (INR 58,888 Crores) funds research through grants and cooperative agreements to more than 2,000 colleges, universities, K-12 school systems, businesses, informal science organisations, and other research organisations throughout USA. The NSF has a Division for Materials Research, of which there is a Biomaterials Group. Of relevance also is the Cellular and Biochemical Engineering (CBE) programme under the Engineering Biology and Health Cluster (Directorate for Engineering). The programme includes Disability and Rehabilitation Engineering, and Engineering of Biomedical Systems. An example of a relevant project grant is $533,338.00 (~INR 4 crores) for research on polymer therapeutics for bone regeneration as next-generation osteoporosis treatments.

The National Institute of Biomedical Imaging and Bioengineering is an organisation of the National Institute of Health, whose mission is to improve health by leading the development and accelerating the application of biomedical technologies. The Division of Discovery Science & Technology (bioengineering) is particularly relevant and the organisation is responsible for research funding.

In Australia, the Medical Research Future Fund (MRFF) was set up by the Australian Government in 2015 (https://www.health.gov.au/initiatives-and-programs/medical-research-future-fund).

From the MRFF fund, $190.8 million (equivalent to 893 crores INR) are being invested in clinician researcher fellowships, in Australia.

The scheme covers key clinical areas and pays for the salaries of the clinicians, if required, and a research support package for 5 years. Under the Australian Government's Medical Research Commercialisation Initiative, delivered by MTPConnect, the $45 million AUD (equivalent to 210.7 crores INR) BioMedTech Horizons programme supports research programmes involving innovative health technologies and medical devices through their journey from discovery towards proof of concept and commercialisation (https://www.mtpconnect.org.au/biomedtechhorizons). Under this specific scheme, the funding for each project can be typically of up to $1 million (4.6 crores INR). As per the National Critical Research Infrastructure initiative of the Australian Government, funding will be provided for research infrastructure needed for conducting health and medical research.

2.5 A Way Forward

Against the backdrop of the significant socio-economic potential, it is important to establish capabilities of indigenous biomaterials and biomedical implants manufacturing. Towards achieving this longstanding need, it is indeed imperative for a national policy to promote establishing of medtech research parks in India. This would change the ecosystem of the highly import-dependent medical devices segment. This includes not only providing resources for medtech start-ups and indigenous manufacturers, but also providing them with support for the commercialisation, intellectual property, and regulatory aspects. The funding for research programmes in this direction is certainly not substantial, when compared to the current status in many developed nations. Furthermore, it is essential for a trained workforce to be in place for the medical parks to be a success. This needs revamping the medical education system as well as introduction of bioengineering programmes in more institutes of national importance.

It can be perceived that translational research needs to have a greater impact in India, so that the outcome can be translated to the commercialisation of indigenously developed products and technologies in the medtech market. This should be pursued without compromising the quality in terms of clinical performance of the devices and implants. The far-reaching impact of implementing a more robust ecosystem can therefore be very significant in terms of the patient's well-being in the world's second largest populous country (India), and this will be a model for other developing nations.

Chapter 3
Scientists at Work in India

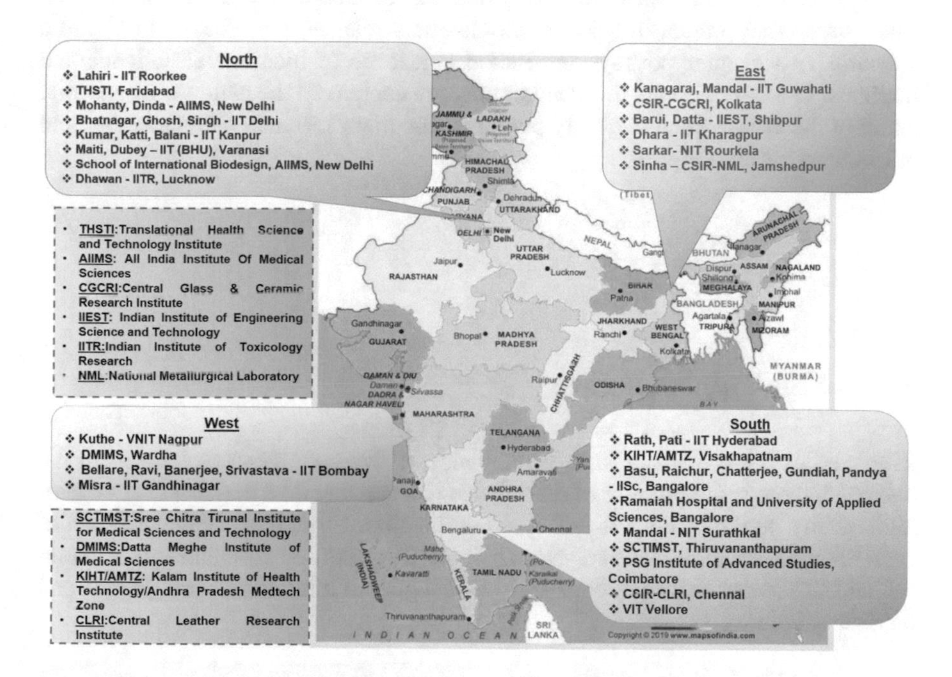

Abstract As breakthroughs in biomaterials science reach a tipping point, a *tour de force* of the diverse research in laboratories of excellence across India can be seen to occur, promising lower healthcare costs and a better future for patients. Who are the scientists, what are they up to and where? The rise of engineering and advances in biology opened up new frontiers in the twentieth century. In the new millennium, as the two started to converge, biomaterials sciences took shape, creating

Disclaimer: The presentation of material and details in maps used in this chapter does not imply the expression of any opinion whatsoever on the part of the Publisher or Author concerning the legal status of any country, area or territory or of its authorities, or concerning the delimitation of its borders. The depiction and use of boundaries, geographic names and related data shown on maps and included in lists, tables, documents, and databases in this chapter are not warranted to be error free nor do they necessarily imply official endorsement or acceptance by the Publisher or Author.

B. Basu, *Biomaterials Science and Implants*,
https://doi.org/10.1007/978-981-15-6918-0_3

an unprecedented link amongst engineers, biologists, clinicians and entrepreneurs. Numerous initiatives are underway at the national and institutional levels in India, with a number of research groups working in academia and national laboratories (see geographical map and groups). However, challenging barriers exist between the bench and bedside; as many research programs are ended at late stages of laboratory-scale development or pre-clinical validation. A few examples from the academia and national laboratories will be provided to highlight the status of the clinically validated indigenous technologies. It is noteworthy that funding and facilitation are required for human clinical trials, an important step in translational biomaterials science. Therefore, significant challenges pave the road to this transformative partnership: how to translate biomaterials and bioengineering concepts to improve human health? Like every other branch of science, biomaterials science, too, does not exist in a vacuum. This chapter portrays the Indian landscape of biomaterials science and implants. The discussion in this chapter may not contain all the national groups, and many of these have been currently pursuing important research themes in the field of biomaterials and implants.

3.1 At the Tipping Point

Biomaterials science stands at the confluence of two contradictory forces. On the one hand, there is the promise of the science that can meet the massive challenge of high-quality public healthcare in India. On the other hand, there is the question of cost-effectiveness that can be solved only by taking world-class scientific work from Indian laboratories to the patient's bedside, keeping in mind time, cost, quality, and regulatory hurdles.

> A handful of research groups are engaged in translational research—a concept that describes scientific investigation, aimed at transferring laboratory discoveries into new methods for diagnosing, preventing and treating diseases, as well as using the knowledge gained to improve healthcare policy.

I will discuss the work of scientists from many disciplines who often work in collaboration across laboratories. Their research activities, funding status, commercialisation efforts, and regulatory compliance reveal the landscape of this frontier science in the country.

3.2 All About Joints: Hips, Knees, and Spines

CSIR-CGCRI scientists, over the last two decades, have developed new manufacturing technology to develop Al_2O_3-based ceramic femoral heads, ZTA-based femoral heads, HA-based coatings for metallic implants, drug-eluting femoral stems, HA/TCP-based granules for the clinical domains of orthopaedic and/or dental surgery. An Al_2O_3-based ceramic femoral head has been developed by CSIR-CGCRI for ceramic-on-PE total hip replacement (THR) and those femoral heads have undergone single centric clinical trials. They patented this technology for manufacturing process for a modular Al_2O_3 based ceramic femoral head, which can be fitted with a standard hip stem, for THR surgery (see Fig. 3.1). The manufacturing technology for depositing a crystalline hydroxyapatite coating on hip stem implants is also being developed in the same CSIR laboratory. To obtain better adherent coatings, the synthesis of hydroxyapatite powders has been tailored to ensure better flowability during plasma spraying.

Most of the biomaterials or implants developed at of CSIR-CGCRI have been validated in at least one single-centre clinical trial.

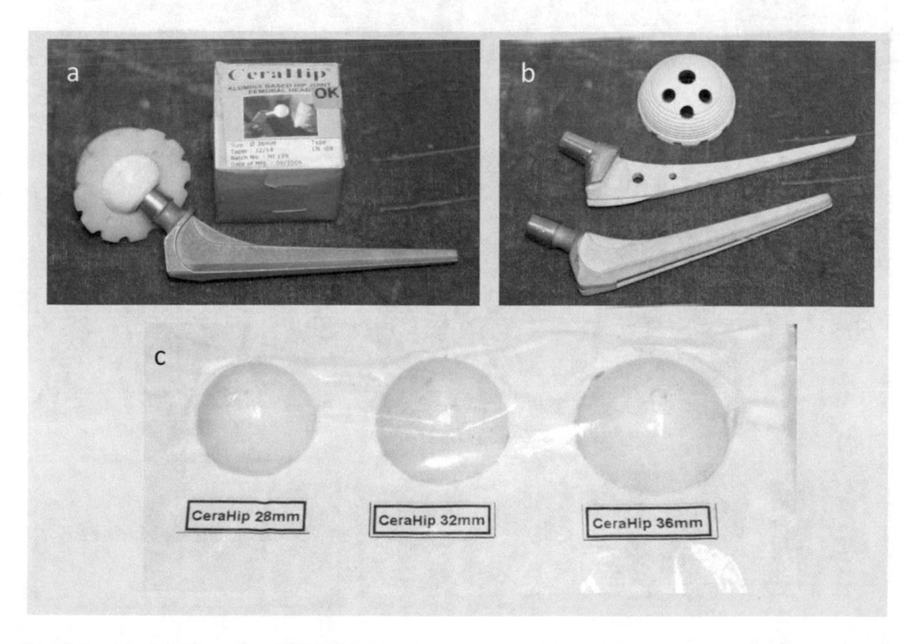

Fig. 3.1. (**a**)–(**c**) Ceramic-on-Polyethylene implants for total hip joint replacement and (**b**) Plasma-sprayed hydroxyapatite coating on metal hip implants, commercialised products developed at CSIR-CGCRI, Kolkata (Images courtesy CSIR-CGCRI, Kolkata)

Life of a metal-on-plastic THA is often limited to 10 years, mainly due to excessive wear debris generation during its operation, leading to aseptic loosening. Machining of an extruded block of UHMWPE leaves some permanent marks on the commercially available liner, which is also one of the reasons for severe wear debris generation. An Al_2O_3-based ceramic femoral head has been developed by CSIR-CGCRI for ceramic-on-PE total hip replacement (THR) and those femoral heads have undergone single centric clinical trials.

The author's research group (earlier at IIT Kanpur and currently at IISc Bangalore) has developed new biomaterials concepts and conducted biocompatibility assessment on implants for musculoskeletal regenerative applications. In particular, their work on three different generations of polymer-ceramic hybrid acetabular liner (total hip replacement, also called total hip arthroplasty or THA), has resulted in new alternative biomaterials options. For acetabular liners, UHMWPE/HDPE blend-based composites reinforced with surface functionalised graphene oxide (GO) were found to effectively enhance mechanical strength, clinically desired wear resistance with less wear debris generation and biocompatibility (see Fig. 3.2). The hybrid composite has shown excellent bone regeneration, when implanted in rabbits' femoral defect for 12 weeks or in segments of tibia over 14 weeks post-implantation.

A new generation of polymer-ceramic hybrid acetabular liner has been developed using three different manufacturing routes by IIT Kanpur, IIT Guwahati, and IISc Bangalore researchers.

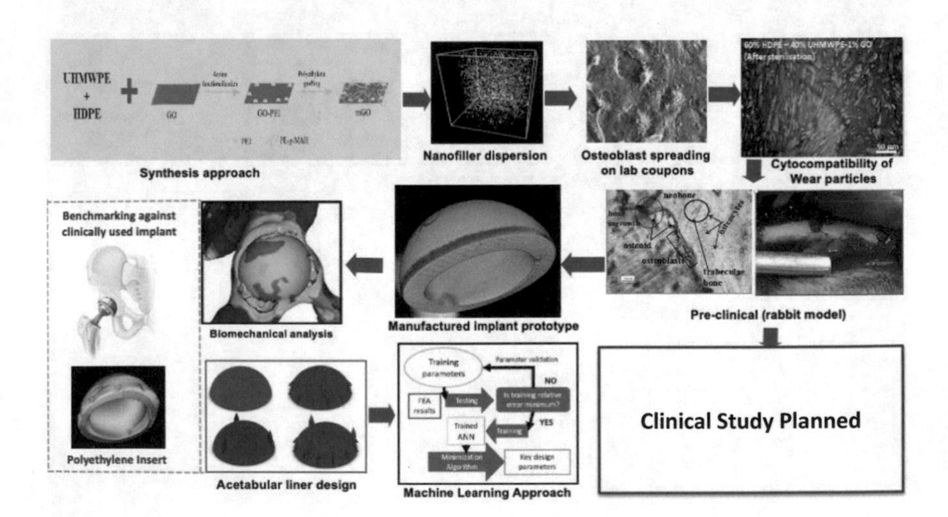

Fig. 3.2 Polymer-Ceramic hybrid acetabular liner: Science-to-Product Prototype (research conducted in author's group at IIT Kanpur and IISc, Bangalore)

In particular, the collaboration between IISc Bangalore (Basu) and IIT Guwahati (Kanagaraj) researchers has led to the development of an acetabular liner of UHMWPE/MWCNT/nHA nanocomposite, with superior wear resistance, better biocompatibility, and high surface finish, leading to improved longevity of the metal-on-plastic prosthetic. Importantly, cost analysis for hybrid polymer-ceramic composite-based acetabular liner implants, shows a 3–4 times lesser cost, when compared to that of the imported ultra-high molecular weight polyethylene (UHMWPE) liners.

Another collaboration between NIT Rourkela (Sarkar) and IISc (Basu) in the field of orthopaedics resulted in patenting a new manufacturing technology for making ceramic femoral heads.

While using an Integrated Computational Materials Engineering (ICME), the team of researchers from NIT Rourkela (Sarkar) and IISc (Basu) judiciously developed process protocols for femoral head prototypes of zirconia-toughened alumina (ZTA) with reliable mechanical properties, including clinically relevant burst strength, together with blood and tissue compatibility, in mouse models.

Based on the outcome of the pre-clinical study, the ceramic femoral head, manufactured using a cost-effective route, has advanced into a human clinical trial on osteoarthritic patients to obtain scientific evidence (clinical and radiological outcomes) of the safety and efficacy in total hip replacement surgery.

Research at Kanagaraj's group at IIT Guwahati is in progress on a multi-axes ankle joint, direct socket fabrication system, knee brace for osteoarthritis patients, reciprocating gait orthosis, and a mid-face external distraction device. In particular, the development of the polycentric knee joint (IITG Knee or SANKALP Knee) has gone through three evolutionary stages, with each having gone through design improvements and additional features to improve its functionality. The first two generations were used by 12 trans-femoral amputees in India. The testing of orthopaedic devices, like total hip or knee replacement implants demand the use of costly equipment, like hip or knee joint simulator, respectively. Like many other scientific equipments, these test facilities are mostly imported and are available in less than five places, including academic institutions and national laboratories.

Kanagaraj and his team at IIT Guwahati developed an ISO standard hip joint motion simulator with four degrees of freedom, for the first time in India. The present version is a modular version of a simulator, which can be modified as a knee joint simulator and intervertebral disc wear simulator.

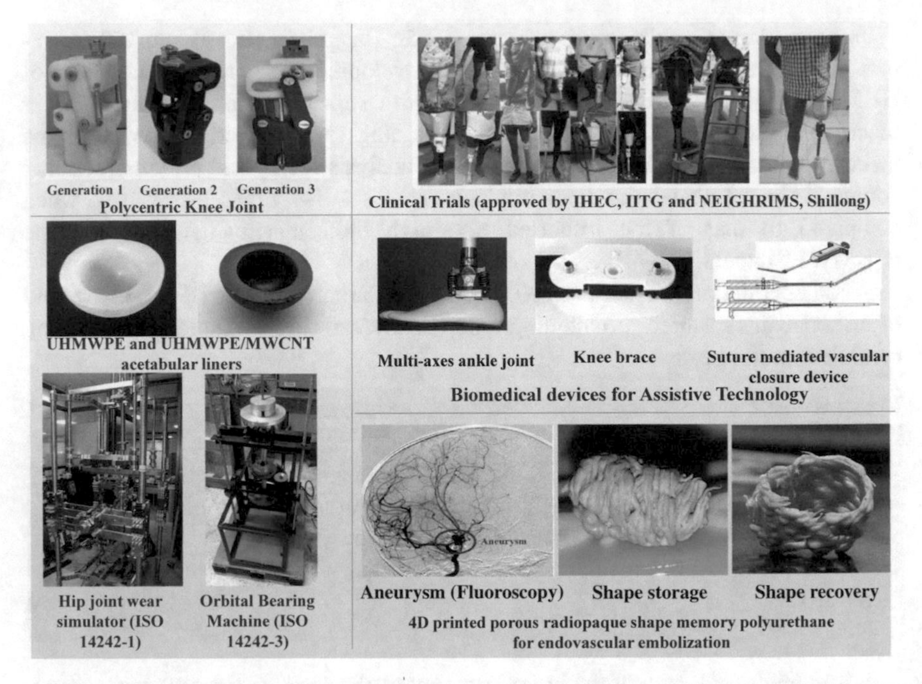

Fig. 3.3 Overview of biomaterials or biomedical engineering research in Kanagaraj's group at IIT Guwahati (Images courtesy S. Kanagaraj)

The past and present research programs at the gait and motion analysis laboratory of IIT Guwahati, under the leadership of S. Kanagaraj, also focus on developing a knee brace for the conservative treatment of osteoarthritis, preservation of residual hearing after fixing the cochlear implants, shape memory polymers for endovascular application, bipolar forceps for electrosurgery, bone marrow aspiration device, mid-face external distraction unit, and many more (see Fig. 3.3). His recent work on the near-net shape manufacturing of a polymeric acetabular liner based on UHMWPE-HA-MWCNT nancocomposites will now enter the pre-clinical validation stage. Many of the biomaterials or biomedical device prototypes are also being tested using in-house facilities, like the hip joint wear simulator, orbital bearing machine, and amputee walking simulator (see Fig. 3.3).

3.3 Bones and Teeth

Novel dental implant designs and their clinical efficacy constitute major challenges in the field of reconstructive dentistry.

Naresh Bhatnagar's research group at IIT Delhi has employed biomechanical analysis, and adapted conventional manufacturing processes, to manufacture metallic dental implants, which were marketed in India in late 2019.

The major highlights of the commercialised technology include: (a) simulation of lobed implant screw; (b) manufacturing of designed implants and laboratory-scale testing, including fatigue testing; (c) animal testing (cadaveric), and (d) limited clinical trial.

In an ongoing research program, the author's research group at IISc, Bangalore, has been working on the development of a complete three piece-dental implant stem, made of implant screw, straight and angulated abutments with hybrid thread profiles, locking screw, healing abutment, and cover screw. The work, conceived in collaboration with a prosthodontist, Dr. Vibha Shetty, Ramaiah University of Applied Sciences, Bangalore, has resulted in an indigenous dental implant system to treat edentulism (see Fig. 3.4). Biomechanical compatibility has been demonstrated by using extensive finite element analysis and chewing simulation studies. The combination of sand blasting and acid etching, together with the introduction of dual ceramic coatings (collaboration with SCTIMST, Thiruvananthapuram) on abutment/screw surfaces with an asymmetric antirotation profile and clinically relevant hybrid thread profile, are considered as inventive components in this new implant system. This system exhibits better primary stability (insertion torque in synthetic

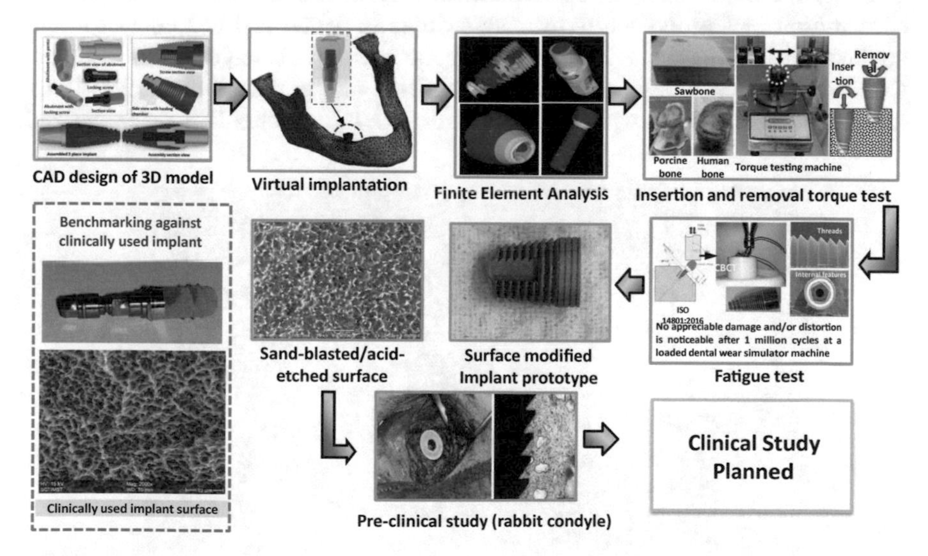

Fig. 3.4 Dental implant system with hybrid threads and antirotational features developed at DBT sponsored translational centre of excellence on biomaterials (IISc-SCTIMST-Ramaiah University of Applied Science collaboration)

and animal bone) and osseointegration in rabbit condyle, when compared to those of the commercial implants.

> The thread design of dental implants and their biomechanical validation to assess periprosthetic bone response in a clinically acceptable manner constitutes one of the major research themes in reconstructive dentistry at IISc, Bangalore.

The technology for depositing crystalline hydroxyapatite (HA) coating on metallic implants using plasma spraying has been established by CSIR for metallic implants for orthopaedic and dental applications. CSIR-CGCRI's coating technology has gone through clinical trials and, technology transfer to interested industries is yet to take place. CSIR- CGCRI has also developed a simple and economical wet chemical synthesis process to make calcium phosphate (CaP)-based granules of different sizes and porosities, for dental and bone filling applications (see Fig. 3.5). This process provides high-purity HA and HA + TCP (biphasic) granules with high crystallinity and good biological properties. Multi-ion-doped HA-based bioceramics, are being developed at CSIR-CGCRI and IISc, Bangalore for accelerated tissue integration with antimicrobial properties.

Bone cement is being used for fixing implants at the hip and knee joints. Using a different bioceramic composition than CASPA, developed by SCTIMST scientists, CSIR-CGCRI researchers formulated an injectable bone cement with patented composition having the ability to release drugs in dental and bone defect filling applications. An injectable biodegradable bone cement is being designed at CSIR-CGCRI. The technology constitutes composition and preparation of self-setting, injectable bone cement, with and without drugs. The process is inexpensive and

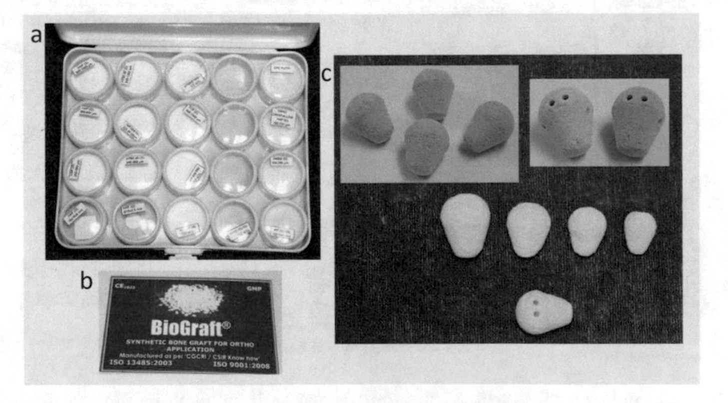

Fig. 3.5 (**a**) Calcium phosphate granules/scaffolds for dental and orthopaedic applications (**b**) BioGraft packaging of final product of (**a**); (**c**) Hydroxyapatite-based orbital implants, commercialised products developed, and clinically tested (Images courtesy CSIR-CGCRI)

produces cement with excellent injectability, setting, and biological properties. Clinical trials are completed and modification of the cement with radiopaque particles to enable post-implantation identification and evaluation is in progress.

The Functional Materials Group at CSIR-NML, led by Arvind Sinha, is among the pioneers in India to initiate research on biomimetic materials (that mimic biological processes) using nanomineralisation processes, akin to matrix-mediated biomineralisation. The group has developed and patented a number of processes for the synthesis of nanosized hydroxyapatite and biphasic powders and nanocomposites of various types: injectable, polymer-hydroxyapatite, polymer biphasic to carbon fibre-reinforced polymer hydroxyapatite.

It is interesting to note, that even in the absence of any biological growth factor, the nanopowders and nanocomposites developed by CSIR-NML exhibit bone healing processes (osteoinduction and osteointegration). This was a collaborative work between CSIR-NML and CSIR-CCMB, Hyderabad. Different polymer-bioceramic nanocomposites developed by CSIR-NML exhibit mechanical properties (compressive strength and modulus) close to those of cancellous bones (spongy bones found at the ends of long bones, in the pelvis, ribs, skull, and the spinal column).

The technology of nanobioceramics has been transferred from CSIR-NML to different industries, namely M/S Eucare Pharmaceuticals, Chennai; M/S Surgiwear Pvt Ltd, Shajhanpur; and M/S IFGL Pvt Ltd, Kolkata. Three products, namely SYBOGRAF, SYBOGRAF PLUS, and SHAP are already in the market.

In collaboration with an Indian industry, CSIR-NML is also involved in the development of the third-generation 'biodegradable metallic implants'. These materials are being designed to support the injured part through the restorative process and are anticipated to degrade within the body, thus avoiding the need for revision surgeries. The mechanical strengths of the currently used biodegradable polymer implants are very poor for use as orthopaedic implants. Therefore, metallic biodegradable alloys (Mg-based, Fe-based, and Zn-based) are currently being developed at CSIR-NML for orthopaedic applications.

In new research programs at CSIR-NML, Mg-based alloys are preferred, because of their comparable elastic modulus (ratio of stress to strain) to that of human bone, eliminating stress shielding and implant loosening issues. In addition, Mg alloys are reported to stimulate the growth of osteocytes and are biocompatible. CSIR-NML has developed a series of Mg-alloys through conventional melting and casting. Although the requisite mechanical properties are achieved in Mg-alloys through post-processing, the degradation rates of the alloys are much higher than those required for biodegradable implants. Therefore, the present focus at CSIR-NML is on achieving desired degradation rates through: (i) alloy design strategies; (ii) surface modification of the developed alloys; (iii) development of suitable biocompatible coatings.

Balani's research group at IIT Kanpur, focuses on hydroxyapatite (HA)-based coatings and scaffolds. The group has used carbon nanotube (CNT) material to significantly toughen HA, while attaining enhanced bioactivity. Further, his group has adapted functionally gradient materials to provide surface bioactivity, while engineering a tougher material for biological applications.

Dhara's group at IIT Kharagpur has developed a green machining approach, as a net-shape forming technique, for fabricating customised products, like ceramic crowns, Ti6Al4V-based dental roots, hybrid mandible/bone plates, etc.

Extensive commercialisation efforts are currently underway through Dhara's own start-up company, Amnivor Medicare Pvt. Ltd, which is focused on development of value-added products for skin, bone, and healing of osteochondral defects (injury to the smooth surface on the end of bones).

The products of Amnivor include wound dressings, hemostats, dental solutions, and various biotechnology products like bioink for 3D printing and 3D cell culture matrices.

Although HA-based porous scaffolds are under development in the last three decades, the naturally derived biomimetic HA, with tunable elastic modulus and strength, together with faster biomineralisation property, remains to be achieved. To address this specific issue, a group of researchers led by Saumen Mandal from NIT Surathkal collaborated with IISc researchers to develop a scalable biomimetic synthesis approach, to obtain submicron HA powders, from marine benthos (Fig. 3.6). Depending on pore-formers, the scaffolds (with a range of porosity up to 51%, with a larger range of pore sizes of up to 70 μm) were fabricated. A combination of moderate compressive strength (12–15 MPa) with an elastic modulus of up to 1.6 GPa was

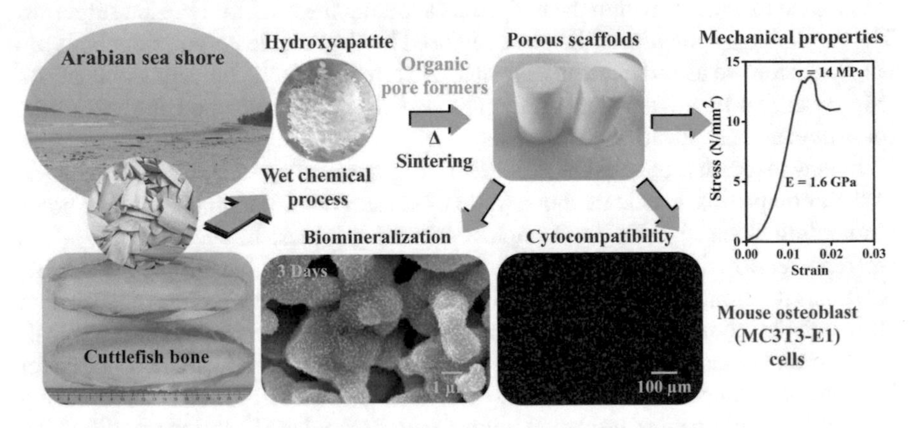

Fig. 3.6 Development of biomaterials from marine benthos (Collaboration between Mandal's group at NIT Surathkal and the author's group at Indian Institute of Science, Bangalore)

obtained with HA, derived from cuttlefish bone, and with wheat flour as the pore-former. Most importantly, this specific HA scaffold supports faster nucleation and growth of the biomineralised apatite layer with full coverage, within three days of incubation in simulated body fluid, as well as better osteoblast cell proliferation, in comparison with those of the synthetic counterpart. The technology of adherent HA/ZrO_2 coatings on metallic substrates, such as titanium alloy and stainless steels, is established at low processing temperatures.

3.4 Cartilage, Bone, Cardiac, and Skin Tissue Engineering

In this subsection, I review advancements in Indian laboratories on translational approaches to replace damaged tissues or whole organs and directions for further development. The research activities of the Biomaterials and Tissue Engineering group of Kaushik Chatterjee at IISc Bangalore can be broadly divided into the following three areas. Firstly, this group has developed 3D scaffolds from novel biodegradable polymers and polymer nanocomposites, incorporating different types of ceramic and carbonaceous nanomaterials to promote tissue generation in the scaffolds. Scaffolds are prepared by various techniques such as 3D printing, porogen leaching, and electrospinning. Secondly, engineering of organotypic tissue models is ongoing.

The most recent research in Chatterjee's group involves studying human cells in 3D scaffolds towards engineering tissues *in vitro*, that mimic the cell response *in vivo* to facilitate the study of cell biology and drug screening in tissue-like environments, including breast tumour, intestinal, and cardiac tissues.

Thirdly, Chatterjee's group is working on additive manufacturing of Ti-alloys and surface engineering techniques for use in the next generation of orthopaedic implants with enhanced mechanical and biological performances.

The research contributions of Dhara's group at IIT Kharagpur are based on a fundamental understanding of materials property in relation to manufacturing processes, customised implant development, cell–material interactions, and *in vivo* biocompatibility. His research has been translated to deployable technologies and commercially viable products through indigenously developed platform technologies. The three focus areas of his group are engineered biomaterials for cost-effective therapeutics towards skin and skeletal tissue regeneration; customised dense, and porous titanium/ceramic implants through novel powder metallurgy-based techniques, such as coagulation casting and plastic dough processing; and simple green approaches for the preparation of antimicrobial, biodegradable radiopaque polymers, and functionalised carbon dots for healthcare applications.

In the domain of engineered biomaterials, Dhara's group has utilized several approaches to reprocess biowaste (fish scale, egg shell, sea shell, placenta, and onion peel) for skin and skeletal tissue regeneration. Dhara has also formulated several bioengineered products based on the combination of Collagen I, Collagen II, and decellularised extracellular matrix. He developed a chitosan citrate gel for osteochondral defect healing. Further, carbon quantum dots-doped calcium phosphate nanorods was demonstrated to facilitate ectopic chondrogenesis via activation of the HIF-α/SOX-9 pathway.

The tissue engineering group at PSG Institute of Advanced Studies, Coimbatore, has been working on scaffolds for damaged human knee meniscus replacement, and to develop techniques that enhance the growth of primary human meniscus cells. The challenge in meniscal tissue engineering is the lack of homogeneity in the meniscus—be it from difference in cell types, collagen arrangements, biphasic behaviour, or geometrical differences. To address such a challenge, the researchers of PSG Institute developed poly-ε-caprolactone/carbon nanofiller scaffolds with high electrical conductivity and moderate roughness which exhibit enhanced cell attachment and proliferation in simulated body fluid and phosphate buffer saline. Poly-ε-caprolactone-based nanocomposite scaffolds with different concentrations of carbon nanofillers (carbon nanofibres, nanographite, and exfoliated graphite) have been studied to investigate the influence of electrical conductivity and biomolecule supplementation for enhanced human meniscal cell attachment, and proliferation. Nanoporous honeycomb garland-like structures are created on the pore walls of these scaffolds. Through biochemical and immunohistochemistry analysis of these scaffolds, in a rabbit model, biocompatibility has been established, in collaboration with PSG Institute of Medical Sciences and Research and the Ortho One Hospitals Group, Coimbatore. Another collaboration is in place with Innov4Sight Health and Biomedical Systems Private Limited, a Bangalore-based company for stem cell research.

While working on bone and neural tissue engineering, PSG Institute of Advanced Studies, Coimbatore, has set up research collaborations with RMIT University and St. Vincent Hospital, Australia, to develop nerve guidance conduits.

A research group at the Central Leather Research Institute (CSIR-CLRI) has developed a novel graphene-metal-amine-reinforced haemostatic and wound dressing biomaterial. The biomaterial has synergistic effects on blood clotting and has antibacterial, wound healing, and absorptive properties. The biomaterial has more fluid absorption capacity and higher mechanical strength than those of conventional dressings.

Neetu Singh's laboratory at IIT Delhi has developed a plasma-induced graft polymerisation method to improve the cell adhesion property of *Bombyx mori* silk fibroin films (SF), which enables the development of silk-based surface coatings with enhanced cell adhesion profile as well as spatial control over cell adhesion. Further, the group has demonstrated that stem cells can be differentiated into bone and cartilage cells on the same scaffold without adding any exogenous enhancers. The control over the release kinetics in the range of a few hours to an extended period of time (temporal control), for simultaneous delivery of signaling molecules has also been achieved. A region-specific activity of biomolecules, by patterning the silk substrates to direct cell adhesion using nanoparticles, has also been investigated. For bone tissue engineering applications, the scaffold has been modified to improve antimicrobial properties to combat biofilm formation of antibiotic resistant bacteria, without compromising its biocompatibility and stem cell differentiation potential.

A 3D scaffold system developed at IIT Delhi, supporting fast and efficient differentiation of stem cells into osteoblasts and chondrocytes, advanced the field of tissue engineering towards creating the complex architecture of a tissue from stem cells, which has been a long-standing problem.

Naresh Bhatnagar's research group at IIT Delhi has developed manufacturing protocols for the fabrication of a commercially competent vascular stent, using PLA/PCL blends. From optimisation of material processing and manufacturing processes to surface properties evaluation and limited cytocompatibility assessment—the research delves into an end-to-end study of immense translational value. An innovative technology was developed allowing for the fabrication of dimensionally accurate stent tubes in a single step process. Surface properties have been improved to induce better cell-substrate interaction.

At IIT Bombay, a number of research groups led by faculty members, including Jayesh Bellare, B. Ravi, Rinti Banerjee and Rohit Srivastava a.o. have contributed significantly to the field of biomaterials and implants. Bellare's work has led to the invention of biodegradable bone graft materials.

Rinti Banerjee's group works on smart, stimuli-responsive biomaterials for drug delivery, point of care diagnostics, medical devices, and technologies for global health. The work in Banerjee's group encompasses the entire spectrum from biomaterial development to *in vivo* evaluation and translation for scale up and commercialisation, working closely with clinical and industry partners. Banerjee's group has developed many commercially viable technologies, seven of which have been licensed to industry partners. This includes smart *in situ* gels for depot intravesical drug delivery in the urinary bladder. This patented Urigel technology focuses on eliminating the barriers to delivery of drugs through the bladder wall while preventing washout by urination.

The Urigel technology and its application for chemotherapeutic and immunotherapy in bladder cancer, have been licensed to Ferring Pharmaceuticals.

3.5 Eyes, Nerves, and Ears

About 40–50% patients are single-eyed, amongst 60–70 lakhs of blind people in India. Significant demand for functional and cosmetic rehabilitation for ailing, single-eyed patients persists in India. A unique product developed by CSIR-CGCRI scientists is the orbital implant for eye surgery.

New design concepts of integrated orbital implants were conceived and manufacturing technology was developed to produce highly porous, yet strong orbital implants, using bioactive hydroxyapatite by CSIR-CGCRI scientists.

The implant comes in different sizes, to suit different patients (see Fig. 3.5). Its high porosity enables tissue ingrowth, providing the natural eye environment that is synchronised with the other natural eye. Clinical trials are complete and technology transfer to interested industries is being pursued.

In the field of peripheral nerve regeneration, the studies at Debrupa Lahiri's group at IIT Roorkee are focused on improving cell adhesion and providing electrical guidance for axonal regeneration. Her group is currently exploring a different, second phase of reinforced polymeric matrices. For skin regeneration, this group has developed a fully biodegradable scaffold, with desired physical and mechanical properties. It has been used in the rapid healing of deep acute and chronic wounds. The scaffold demonstrates the capability to heal severe burn wounds, to prevent bacterial infection, to reduce inflammation, and to regenerate burnt tissues. The researchers at IIT Kanpur (Ashok Kumar's group) have been working on the repair and regeneration of damaged nerve tissues. With strategies for bone regeneration and nerve guidance in place, Kumar's group is undertaking the challenging problem of vertebral disc and spinal cord injuries.

Limited efforts have been invested in India to improve the performance of cochlear implants in the domain of ENT surgery.

IIT Guwahati researchers, led by S. Kanagaraj together with IISc researchers and ENT surgeons of North Eastern Indira Gandhi Regional Institute of Health and Medical Sciences, Shillong, have been conducting interdisciplinary research to improve the performance of conventional cochlear implants.

The team is working on a cerium-based electrode material, that can scavenge excess reactive oxygen molecules in order to preserve residual hearing after cochlear implant fixation. Most commonly used CI electrodes have the outer insulating layer made up of silicones. Some studies have shown that the silicone layer can be used as a drug delivery medium in the location of the CI. In their unpublished research, Kanagaraj's group investigated the pathways to control the ROS by local antioxidant delivery to preserve the residual hair cells, particularly by coating the electrode of the CI with nanoceria or silicone-nanoceria composite. This is important as medium to long-term local delivery of antioxidants remains a challenge.

Nanoceria, an antioxidant material having high oxygen storage capacity and unique redox property of changing the oxidation state between Ce^{3+} and Ce^{4+}, based on the environmental conditions, was used as a catalyst and radical scavenger. The use of nanoceria can potentially reduce the oxidative damage of sensory hair cells. In this research program, the cytocompatibility of the coated electrodes was investigated using hair cell/neuronal cell lines and also primary hair cells isolated from murine organ of Corti. After finalising suitable composite properties and concentration/dosage, the biocompatibility of nanoceria-coated CI electrode and the silicone-nanoceria composite-insulated CI are currently being investigated in murine models, with a particular focus on evaluating the systemic and local inflammation and hair cell damage in the implanted animals.

3.6 Antibacterial Biomaterials to Combat Implant-Associated Infections

Infection is one of the major reasons, which causes implant failure. Globally, implant failure from infections stands at about 2–7%, over the implant's lifetime. The gram-positive bacteria, *Staphylococcus aureus,* has been found to be extensively involved in implant-related infections. More than 30% of failures or rejection of orthopaedic implants are directly linked to post-surgical infection. Aseptic loosening and periprosthetic bacterial infection (around an implant) are the prime causes. The existing remedy lies with antibiotics, but high doses cause unnecessary whole-body exposure and associated problems, namely antibiotic-resistance. To avoid this efficiently, on-site delivery of antibiotics, post-surgery, is required. It is challenging to modify metallic implants to release drug for a few weeks, without affecting their structural, mechanical, and biological attributes. At present, no such metallic implants are commercially available.

To address this problem, the research group of Debrupa Lahiri and Swati Haldar at IIT Roorkee is also designing systems for sustained drug release for currently used orthopaedic implants. The group has been avidly engaged in exploring biodegradable, physiology-friendly alloys, and composites with mechanical properties, similar to that of human bone, to circumvent revision surgeries, stress shielding, and bone weakening issues. Also, magnesium–zinc (Mg-Zn) alloys and composites, in combination with hydroxyapatite (HA), are being investigated with encouraging results. Since magnesium has a high corrosion rate under physiological conditions, the group's research at IIT Roorkee focusses on its corrosion resistance, while improving its mechanical integrity.

The need for more efficient drug delivery strategies are leading to more research on sophisticated drug-releasing implants, as outstanding alternatives to conventional clinical therapies. Recent clinical trials have demonstrated that this technology can improve both quality of life and life expectancy of patients. Further, the surface needs to be optimised to make the drug-releasing period longer.

At IIT Roorkee, Lahiri's research group has designed HA-coated metallic orthopaedic implants, with the surface being modified to release drug, post-implantation, for around three weeks.

In addition, they are also exploring a drug-releasing matrix to reduce friction, as well as, the wear and tear of implants at the joints.

Balani's group at IIT Kanpur has also assessed the adhesion strength of *S. aureus* gram-positive bacteria on various biosurfaces (such as polymeric 'ultra-high molecular weight polyethylene (UHMWPE)', ceramic 'HA', and metallic 'stainless steel and Ti-6Al-4V' real-life bone-implant substrates). They have utilised molecular dynamic simulation to visualise adhesion events on these biosurfaces and quantified the invariant long-range forces, that combine with short-range forces to allow surface proteins of bacteria to interact.

The protein-unfolding events, visualising detachment of bacteria (correlated with experimental de-adhesion events), are being shown for the first time by Balani's research group at IIT Kanpur.

Balani's recent work on constructing porous HA-based scaffolds, with antibacterial and anti-oxidant efficacy (with ZnO and CeO_2 addition, respectively), is aimed towards attaining multifunctional capability for bone-scaffolds. The aspects of incorporating porosity at multi-length scales in HA-based matrix are being engineered (nanoporosity for enhancing protein-material interaction at molecular length scale, micro-texture for assisting directional-cell growth, and bulk-porosity for ensuring vascularisation effects for expedited healing). Additionally, his work on polymeric-liners for hip joints—where, his group has utilised Al_2O_3 reinforcements and carbon nanotubes to provide enhanced tribological resistance to UHMWPE—has gained attention. The multi-length scale tribological aspects highlight how the ceramic reinforcements show enhanced scratch resistance, whereas metallic reinforcement provides enhanced fretting resistance.

A group of researchers at VIT Vellore explored the feasibility of RF plasma treatment in providing the necessary sterility to the implant. They have studied the effect of RF cold plasma on titanium and stainless steel surgical implants—Ti-6Al-4V (Ti64) and 316L SS—to prevent bacterial adhesion. A stable antibacterial activity was obtained in seven days post-plasma treatment.

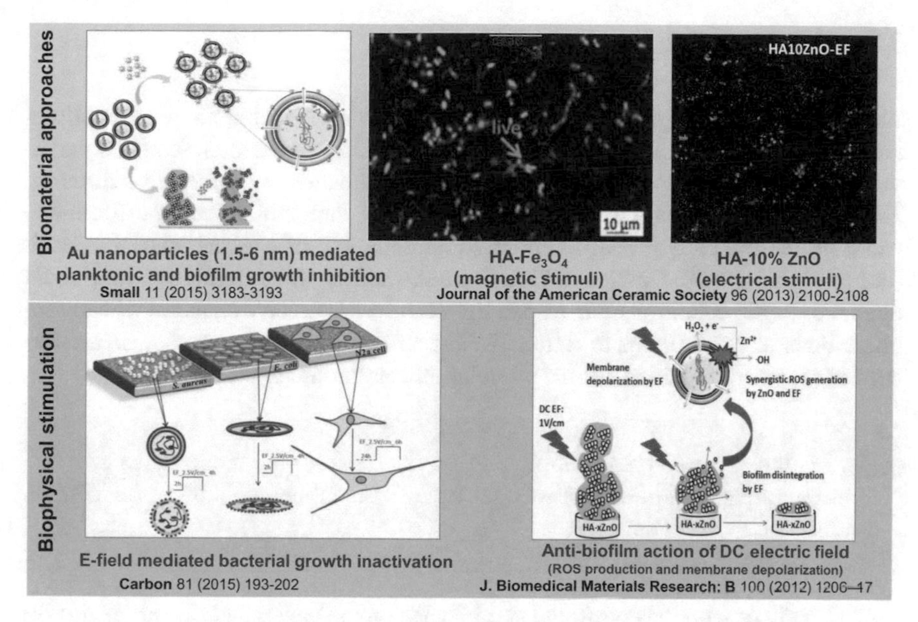

Fig. 3.7 Antimicrobial strategies to prevent implant-associated bacterial infection (summary of author's research at IIT Kanpur and at IISc, Bangalore)

At the author's group at IISc, both biomaterial and biophysical antibacterial approaches have been successfully used. In the context of antimicrobial biomaterials, ultrasmall gold nanoparticles (0.8–1.4 nm) and hydroxyapatite composites reinforced with Fe_3O_4 and ZnO were designed to mitigate MRSA infections. Furthermore, adjuvant treatment approaches, such as external biophysical stimulation in the form of electric and magnetic fields, were demonstrated to enhance the bactericidal properties of the developed antimicrobial biomaterials (Fig. 3.7).

Some of the important scientific discoveries from India in the Bioengineering domain include the synergistic interaction between magnetoactive/electroactive properties of biomaterials and magnetic/electric stimuli (respectively) towards bacteriostatic/bactericidal properties, *in vitro* (author's research at IIT Kanpur and IISc, Bangalore).

3.7 Regenerative Engineering

The author's research group at IISc, Bangalore, has investigated biophysical stimulation of cell functionality on engineered biomaterials. The research focussed particularly on stem cell differentiation for cardiac and/or neurodegenerative diseases. Their research has shown that electrical stimuli can inhibit cell proliferation, while promoting early cell differentiation on a number of engineered biomaterials (doped PANI, PVDF-CNT, HA-$CaTiO_3$, etc.) and in microfluidic devices under dynamic culture conditions. His research team solved Poisson's equation with appropriate boundary conditions to rationalise the role of bioelectric stresses on cellular deformation and identified critical stimulation conditions for electroporation.

> The author's research group unraveled that the intermittent delivery of magnetic field stimuli drives osteogenesis of human mesenchymal stem cells on magnetoactive biomaterials.

This body of work has profound significance on the bioengineering strategies for neural, bone, and cardiovascular regenerative applications.

Hardik Pandya's group (Biomedical and Electronic Engineering Systems Laboratory and Advanced Microsystems and Biomedical Devices Facility) at the Department of Electronic Systems Engineering in IISc Bangalore works closely with a team of clinicians from India and abroad, to develop systems and solutions tailored to address unmet clinical needs. The key focus areas are label-free phenotyping of cancer tissues, cytology-based point-of-care screening for cancer, development of novel steerable catheters for atrial fibrillation and tracheal stenosis, platforms for rapid on-chip antibiotic susceptibility testing, neurophysiological interventions to evaluate mental wellness, development of the phenotypic brain atlas model, novel futuristic bioresorbable sensors for the recording of biopotentials, etc. Each research problem is addressed by an interdisciplinary team with constant feedback and inputs from clinicians. The team leverages advances in microfabrication, electronic systems engineering, additive manufacturing, and machine learning, while combining with established techniques and protocols in biology and medicine to develop clinically relevant solutions.

A number of researchers in the laboratory of Mohanty at AIIMS, New Delhi; and Barui and Datta at IIEST, Shibpur are working towards developing an insightful understanding of the functionality of different types of stem cells on tissue engineered scaffolds.

Ashok Kumar's research group at IIT Kanpur focuses on the development of the new generation of cryogel biomaterials, smart polymeric materials, and nanomaterials towards bone tissue engineering applications. His research group has developed an indigenous cryogel technology, used for stem cell separations, tissue engineering, and regenerative medicine, such as biomedical devices (for instance, an extracorporeal bioartificial liver device).

In another direction, the research team of Ashok Kumar at IIT Kanpur has utilised temperature-responsive smart polymeric material in self-assembling biomolecules and cells for their enhanced utility in regenerative medicine. Further, the laboratory focuses on natural and synthetic nanomaterials for their therapeutic applications in different healthcare problems: cartilage, bone, skin, cardiac, liver, and neural tissue engineering applications. In liver diseases, the focus of Kumar's group is to explore different treatment possibilities to bridge the gap between liver transplantation and regeneration.

Ashok Kumar's group at IIT Kanpur recently patented an integrated bioartificial liver device, with the aim of up-scaling the design for pre-clinical trials in larger animal models. The group is working towards exploiting the therapeutic aspects of exosomes (nanosized vesicles that can travel between cells) in liver tissue engineering.

Further, these materials are being explored for musculoskeletal defect healing and bone regeneration. For example, pre-clinical trials on rat and rabbit in tibia and cranial bone defect models have validated various therapies and approaches using developed scaffolds and fillers based on cryogels, 3D printed scaffolds, and injectable bone cements. Kumar's group has successfully fabricated antioxidant polymeric materials and developed them into 3D scaffolds, which have shown promising applications for developing cardiac patches for treatment of myocardial infarction.

At the Ceramic Engineering Department in IIT (BHU), Varanasi, Dubey has established a state-of-the-art biomaterials laboratory to establish research programs in the broad area of 'Piezoelectric Biomaterials'. In particular, his group is focusing on developing an understanding of the combined action of electrostatic and dynamic electrical stimulation on antibacterial and cellular responses of piezoelectric biomaterials, such as perovskite niobates, silicates, and titanates (see Fig. 3.8). Towards developing a theoretical understanding of these processes, his research laboratory is involved in computational analyses of the functional response of electromagnetic equivalents of mammalian and bacterial cells.

Among the piezoelectric perovskites, Dubey's research group demonstrated the biocompatibility of sodium potassium niobate [$Na_xK_{1-x}NbO_3$ ($x = 0.2 - 0.8$), doped with Li]-based functionally graded composites, using a series of electro-mechanical, impedance spectroscopy, and *in vitro* cellular studies.

In addition to the inherent piezoelectric materials, Dubey's research group at IIT(BHU), Varanasi, induced ferroelectricity, which is similar to that of cortical bone, even in non-piezoelectric hydroxyapatite, via the concept of development of a functionally graded material with sodium potassium niobate. Here, a reasonable

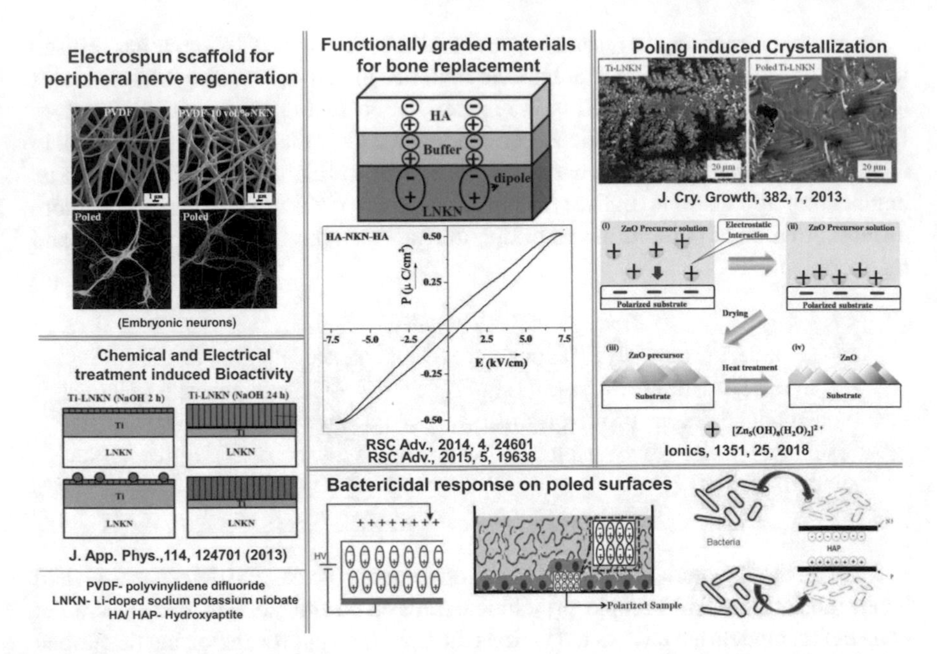

Fig. 3.8 Piezoelectric Biomaterials Research at IIT (BHU), Varanasi (Image courtesy Ashutosh K. Dubey)

combination of bone-mimicking dielectric and electrical response such as, dielectric constant (38), AC conductivity [~10^{-9}/(ohm cm)], piezoelectric strain coefficient (d_{33}-4.2 pC/N), electromechanical coupling coefficient (0.17), mechanical quality factor (81), and remnant polarisation (0.06 μC/cm^2) has been achieved. His group further developed PVDF-NKN-based electrospun scaffolds and investigated the modulation of the neural cell functionality on those scaffolds. The pre-clinical study on the *in vitro*-tested biomaterial systems has been planned.

3.8 3D (Bio) Printing of Biomaterials, Tissues, and Organs

In the field of additive manufacturing, laser or electron beam–based 3D printing is widely investigated for biomedical applications. In a recent work, the author's research group at IISc Bangalore has developed a quantitative understanding of the process physics of 3D binderjet printing, which allows processing at physiologically relevant conditions (Fig. 3.9). This work proposed the formulation of an *in situ* polymerisable acrylic binder for printing implantable metallic biomaterials. The author's team has insightfully extended the Washburn model to rationalise the timescale of binder infiltration to the powder bed with respect to polymerisation kinetics. Based on this scientific study, his research group collaborated with a team of neurosurgeons from Ramaiah Memorial Hospital, Bangalore to conduct a clinical study on the use of

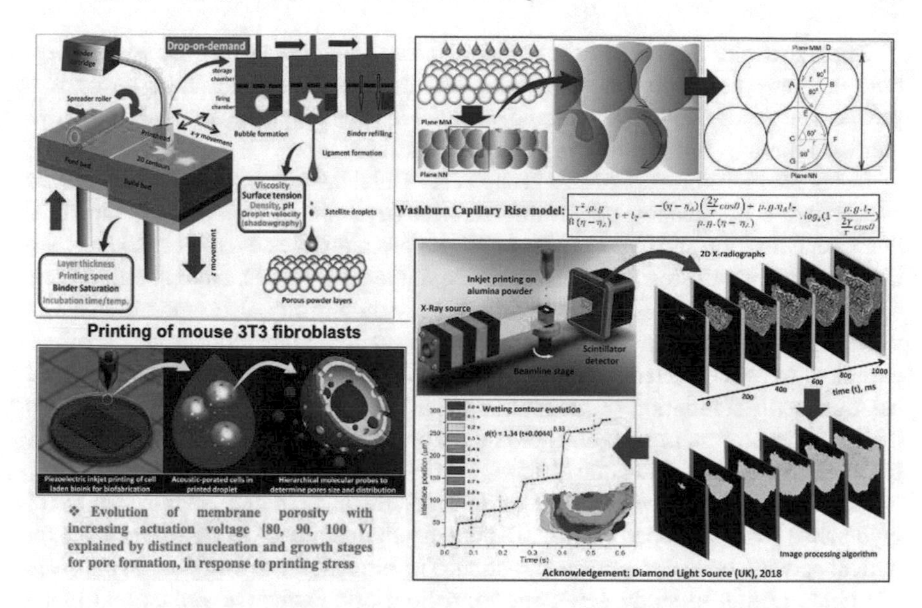

Fig. 3.9 Schematic of the Salient Scientific Research on 3D Binderjet Printing (inkjet printing) of Implantable Biomaterials at IISc, Bangalore

patient-specific cranioplasts to treat decompressive craniectomy in human subjects. A pilot study is currently under progress in another hospital in India.

> The technology of the clinically tested cranial prosthesis involves an integrated methodology, comprising advanced image processing of patient's CT scans, 3D inkjet printing of patient-specific cranium models, and conventional fabrication of acrylic bone flap and sterilisation.

The research group of Sourabh Ghosh at IIT Delhi is working in the interdisciplinary areas of biomaterials science, developmental biology, and cell signaling. The group has developed an *in vitro* model of mesenchymal condensation during chondrogenic development. Mesenchymal condensation is a pre-requisite of chondrogenesis during embryonic development. The group is trying to understand how subtle variation in morphology and stiffness of nanofibrous silk protein matrices can induce aggregation of human mesenchymal stem cells, mimicking early stage chondrogenesis.

> In the field of 3D bioprinting, Ghosh's research group at IIT Delhi developed anatomically relevant, phenotypically stable cartilage constructs, using silk-gelatin bioink.

These scaffolds exhibited close similarities in the signaling pathways with embryonic cartilage development, suggesting that the engineered cartilage tissue is a prospective tissue equivalent with the potential of providing the essential instructive elements for activating pathways of organogenesis.

Using 3D bioprinting, Ghosh's group at IIT Delhi has tried to recapitulate *in vivo* endochondral ossification-based limb skeletal development, specifically chondrogenic condensation and hypertrophic cartilagenous template development. His group identified the involvement of IHH signaling, indicative of the development of bony collar by perichondral ossification. Ghosh's research group also reported that silk fibroin protein activates Wnt-signaling pathways to trigger chondrogenic differentiation, as well as Hedgehog and parathyroid signaling pathways, to modulate osteogenic differentiation of progenitor cells. The group has further extended this insight to develop an *in vitro* organoid model of the dermal papilla of the human hair follicle and a human skin equivalent.

Researchers under the guidance of Subha Narayan Rath and Falguni Pati at IIT Hyderabad have developed a novel 3D-printed polymer bioactive glass composite for patient-specific bone defect reconstruction. The biomaterial is currently at the stage of animal testing to study new bone formation, osteoinductive ability, and blood vessel infiltration. Rath used 3D bioprinting methods to fabricate blood vessels, bone, and cartilage. To achieve this, mesenchymal stem cells (MSCs) are isolated from primary sources, like adipose tissue and umbilical cord and used in a number of bio-inks by 3D bioprinting methods. De-cellularisation tissue-derived bioink is promising to lead a new generation of bioinks for tissue-specific induction, wherein Pati's group is associated for fabrication of the tissues.

Pati's group, along with a team of researchers from the Asian Institute of Gastroenterology (AIG, Hyderabad), is currently working on developing a macro-encapsulation device for allogenic islet cell transplantation for diabetic patients. In addition, his group uses microfluidic chips to test MSC differentiation, cancer stem cell growth, and subsequent drug testing in lab-on-chip devices.

Silk fibroin (SF)-polyvinyl alcohol (PVA)-blended electrospun nanofibrous scaffolds have been developed to investigate the attachment and proliferation of primary human meniscal cells at PSG Institute of Advanced Studies, Coimbatore. A blend of silk: PVA at 3:1 ratio is suitable for meniscus cell proliferation, when compared to pure SF-PVA nanofibres. Improved compressive and dynamic mechanical properties, similar to those of the native human meniscus, were measured. The *in vivo* biocompatibility study confirmed that the scaffolds were able to induce angiogenesis with minimal inflammatory response. Future translational research being planned at PSG Institute of Advanced Studies include a large animal trial in a pig model for 3D bioprinted scaffolds, and meniscus reconstruction *in vitro* studies in a designed bioreactor.

Biman Mandal's group at IIT Guwahati have employed conventional and latest 3D bioprinting techniques to intricately mimic the architecture of organs/tissues to understand the underlying cell-material crosstalk and its role in tissue regeneration.

Mandal's laboratory specifically focuses on recreating functional tissues, organs, and implants using naturally derived biodegradable biomaterials, i.e. Indian endemic silk in combination with stem cells. Mandal's group has developed a number of prototypes, which are in various phases of pre-clinical/clinical validation. These include smart wound dressings for diabetic foot ulcers, skin grafts for burn injuries, vascular grafts for bypass surgery, vascularised bone grafts as orthopaedic implants, a beating cardiac patch for myocardial infarction, a bioartificial pancreas releasing insulin for type-1 diabetes, 3D-printed intervertebral disc and knee meniscus as orthopaedic grafts and minimally invasive anti-cancer drug-eluting injectable gels for cancer treatment. Similarly, the group has developed multiple *in vitro* human disease models, which can contribute to industry in high throughput drug screening and drug development.

3.9 National Institutes of Importance and Centres of Excellence

3.9.1 *Sree Chitra Tirunal Institute for Medical Sciences and Technology, Thiruvananthapuram*

The Sree Chitra Tirunal Institute for Medical Sciences and Technology (SCTIMST) is an institute of national importance under the Department of Science and Technology, Government of India. In the last few decades, several medical device technologies developed here have been transferred to industrial partners for manufacture and commercialisation—including the Chitra heart valve, a process for preparation of extracellular matrix scaffolds from mammalian cholecyst/jejunum/urinary bladder, a rapid urinary tract infection diagnostic kit with antibiotic sensitivity test, a fibrin sealant, a vein viewer, and warmers of IV/blood fluid (see Fig. 3.10).

A team of biomaterials researchers at SCTIMST has developed a new-generation bioactive bone cement, named CASPA. This is a self-setting formulation, containing osteoconductive inorganic phases, that leads to faster healing of bone defects. Bioactive injectable cements are now highly sought-after materials in specialties like orthopaedics, maxillofacial, and spinal specialties, because of the potential advantages of minimally invasive surgery and drug delivery, which can be applied for bone repair and treatment, for example.

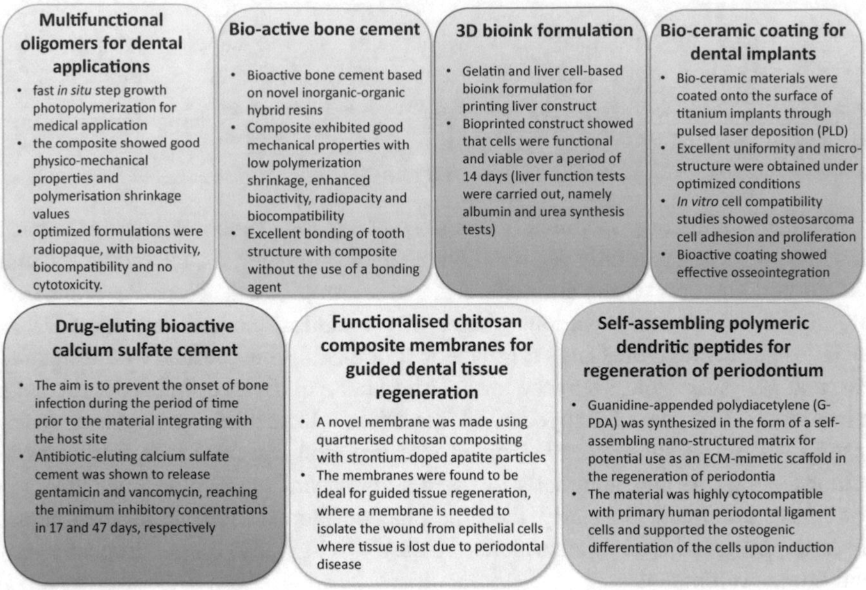

Multifunctional oligomers for dental applications
- fast *in situ* step growth photopolymerization for medical application
- the composite showed good physico-mechanical properties and polymerisation shrinkage values
- optimized formulations were radiopaque, with bioactivity, biocompatibility and no cytotoxicity.

Bio-active bone cement
- Bioactive bone cement based on novel inorganic-organic hybrid resins
- Composite exhibited good mechanical properties with low polymerization shrinkage, enhanced bioactivity, radiopacity and biocompatibility
- Excellent bonding of tooth structure with composite without the use of a bonding agent

3D bioink formulation
- Gelatin and liver cell-based bioink formulation for printing liver construct
- Bioprinted construct showed that cells were functional and viable over a period of 14 days (liver function tests were carried out, namely albumin and urea synthesis tests)

Bio-ceramic coating for dental implants
- Bio-ceramic materials were coated onto the surface of titanium implants through pulsed laser deposition (PLD)
- Excellent uniformity and micro-structure were obtained under optimized conditions
- *In vitro* cell compatibility studies showed osteosarcoma cell adhesion and proliferation
- Bioactive coating showed effective osseointegration

Drug-eluting bioactive calcium sulfate cement
- The aim is to prevent the onset of bone infection during the period of time prior to the material integrating with the host site
- Antibiotic-eluting calcium sulfate cement was shown to release gentamicin and vancomycin, reaching the minimum inhibitory concentrations in 17 and 47 days, respectively

Functionalised chitosan composite membranes for guided dental tissue regeneration
- A novel membrane was made using quartnerised chitosan compositing with strontium-doped apatite particles
- The membranes were found to be ideal for guided tissue regeneration, where a membrane is needed to isolate the wound from epithelial cells where tissue is lost due to periodontal disease

Self-assembling polymeric dendritic peptides for regeneration of periodontium
- Guanidine-appended polydiacetylene (G-PDA) was synthesized in the form of a self-assembling nano-structured matrix for potential use as an ECM-mimetic scaffold in the regeneration of periodontia
- The material was highly cytocompatible with primary human periodontal ligament cells and supported the osteogenic differentiation of the cells upon induction

Fig. 3.10 Selected Technologies from Sree Chitra Tirunal Institute for Medical Sciences and Technology

> The essential validation of CASPA, as per the regulatory requirements, is complete—including pre-clinical studies, while human clinical trials are under progress.

Commercialisation efforts are effectively guided by TIMed, a Technology Business Incubator for Medical Devices and Biomaterials at SCTIMST. In 2017–2018, TIMed supported seven resident and two virtual incubatees in the healthcare domain and was selected for the NIDHI Seed Funding Scheme of DST. This has enabled TIMed to provide seed funds to its incubatees. The group launched the Social Innovation Immersion Programme in the areas of ageing and health with BIRAC support, and four TIMed Innovation Fellows were selected for the scheme.

According to the Institute's Vision 2030: Perspective Plan, a 350-crore project is envisaged for developing biomedical devices, particularly those related to artificial organ development and bioink, orthotics, robotics, and *in vitro* diagnostic devices. SCTIMST aims to develop and transfer 40 innovative medical technologies to industry and take another 20 to advanced stages of development by 2030. Medspark, a medical devices park, is planned to be set up on the premises of the Bio 360 Life Science Park, being developed by the Kerala State Industrial Development Corporation at Thonnakal in Thiruvananthapuram. It is a 180-crore project, funded with financial assistance from the Government of India. The aim of the Park is to create an environment for research and development in the area of medical devices and biomaterials, including testing and evaluation, manufacturing support, technology innovation, and knowledge dissemination.

3.9.2 *School of International Biodesign, All India Institute of Medical Sciences, New Delhi*

The School of International Biodesign (SIB) was built on the success of the Stanford India Biodesign Program. SIB is implemented by DBT, at All India Institute of Medical Sciences (AIIMS), New Delhi and IIT Delhi, in collaboration with QUT Australia and Hiroshima University, Japan. Since the inception of the program, DBT engaged the Biotechnology Consortium of India Limited (BCIL) to manage the techno-legal activities of this program. The aim of this programme is to train the next generation of medical technology innovators in India. This includes idea generation through clinical immersion, need finding, need filtration at AIIMS, and prototype development at IIT, along with IP generation, with the help of BCIL.

More than 50 prototypes have been developed so far, which have been further refined, validated internationally, and tested in preclinical and clinical trials. This has led to the development of over 30 medical devices, which were possible due to the efforts of young innovators, i.e. 60 fellows and over 52 interns. Fifteen technologies have been transferred, and 12 medical technology start-ups have been set up by the fellows trained under this programme in sync with the 'Start-Up' India Program. After completing the programme, many SIB fellows continued their profession in the medtech field, with 38% in medtech industry and around 44% in start-ups.

3.9.3 *Biomedical Engineering and Technology Incubation Centre (BETiC), IIT Bombay*

Based at IIT Bombay, BETiC is an inter-disciplinary multi-institutional initiative for medical device innovation, headed by B. Ravi. Established in 2014, with support from the government of Maharashtra; it comprises a network of 14 engineering and medical institutes across the state.

> The BETiC team has developed 50 medical devices as of 2019, and licensed 16 of them to start-up companies or industry for mass production.

BETiC started functioning from the OrthoCAD laboratory in IIT Bombay in early 2015, followed by two satellite centres at Visvesvaraya National Institute of Technology, Nagpur, and College of Engineering, Pune. The centre organises a Medical

Device Hackathon (MEDHA) in July and a week-long Medical Device Innovation Camp (MEDIC) in September.

The CAD-CAM centre at Visvesvaraya National Institute of Technology (VNIT) Nagpur is home to innovations, under the leadership of A. M. Kuthe. The group has worked on fabrication of customised metallic implants and customised scaffold fabrication using a bioplotter for bone tissue regeneration, whose biocompatibility has been established using animal tests. Until now, around 11 clinical studies are successfully completed, with implants and scaffolds fabricated at the CAD-CAM centre, in diverse areas such as ridge augmentation, sinus lift, socket preservation and guided bone regeneration. Kuthe's group has also developed a non-invasive and anesthesia-free medical device for diagnosis of glaucoma. It is a device to measure intraocular pressure (IOP) through the eyelid. The main advantage offered by this method is an increase in patient's comfort and prevention of corneal infection. The group has conducted many clinical studies with clinicians from Datta Meghe Institute of Medical Sciences, Wardha.

3.9.4 Multi-institutional National Research Programs

IISc, Bangalore, is currently leading the Translational Centre on Biomaterials for Orthopaedic and Dental Applications. The vision is to conduct innovative research that will lead to biomaterials-based solutions in orthopaedics and dentistry. The mission of this centre is to pursue transformative research that will usher in biomedical device development through collaborative efforts of academia, and national laboratories with intensive interactive inputs from clinicians and industries. With the involvement of 43 researchers from multiple disciplines (including bioceramics, biopolymers, biomechanics, mechanical engineering, dentistry, orthopaedic surgery, and histopathology), and 27 young researchers, this centre is currently the largest biomaterials-centric centre of excellence in India. The participants are from CSIR-CGCRI, Kolkata; CIPET, Chennai; IIT Kanpur, NIT Rourkela, SCTIMST, Thiruvananthapuram; IIEST Shibpur; and Ramaiah University of Applied Sciences, Bangalore (Fig. 3.11).

DBT has funded bioengineering and biodesign initiatives at IISc, Bangalore, in two phases. This programme is currently at the Phase II stage. The broad objective of the programme is to strengthen the interdisciplinary basic and translational research programs, by deep engagement with biologists and clinicians from clinical research institutions, including Christian Medical College (Vellore), Mazumder-Shaw Centre for Translational Research (Bangalore), St. John's Medical College and Research Foundation (Bangalore), National Institute of Mental Health and Neurosciences (Bangalore), and a few hospitals in Bangalore. Other objectives include training of MD degree holders to be biodesign innovators and continuing to foster the bioengineering PhD program, by providing research facilities as well as academic monitoring.

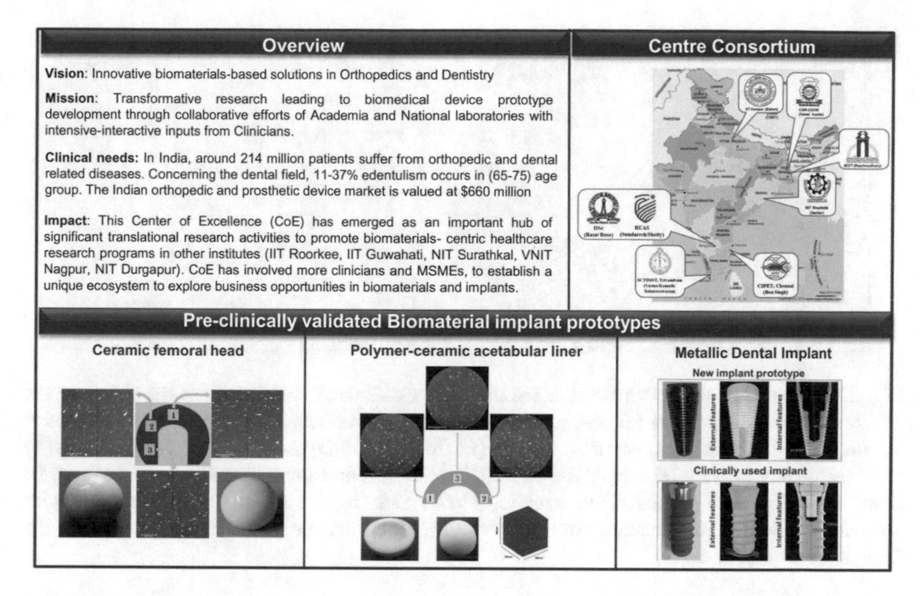

Fig. 3.11 Overview of the Translational research activities of a national center of excellence on Biomaterials for Orthopaedic and Dental Applications, currently being coordinated by the author at IISc, Bangalore

3.9.5 Multi-institutional International Translational Research Programs

Here, I present brief details of some of the recently concluded and ongoing multidisciplinary translational research programs, wherein the author has taken the lead role. The Indo–US Public–Private Networked R&D Centre on Biomaterials for Healthcare involved 25 Indian and US young researchers (http://www.iitk.ac.in/indo_us_biomaterials/). The author's research group, then at IIT Kanpur and Brown University, led this bilateral programme during 2008–2011. The outcome led to the development of biomaterials for orthopaedics, corneal, dermal, cartilage, and cardiovascular tissue engineering applications. Notable achievements of this centre were: (a) genotoxicity of nanobioceramic composites; (b) HA-based electroconductive piezobiocomposites; (c) development of polymer-based scaffold materials for cartilage tissue engineering application; (d) injection molding of polymer-ceramic hybrid biocomposites; and (e) investigating a CAD/CAM-based manufacturing route as well as 3D printing route to fabricate complex-shaped implant materials.

Clinically, osteoporosis is one of the most common health risks faced by the vast majority of ageing populations across the world, especially post-menopausal women. The occurrence of this systemic skeletal disease eventually makes the bone weak, porous, and fragile. The survival of implants in such patients is very low, as osteoporosis affects the process of osseointegration, and patients otherwise have to undergo revision surgery. This unmet clinical need has driven another international

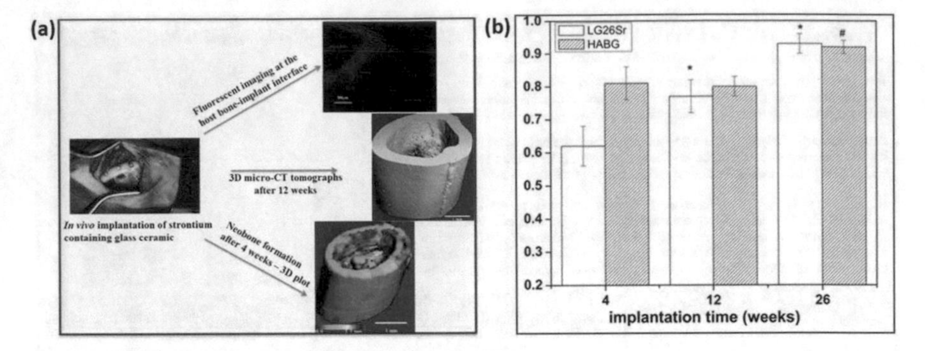

Fig. 3.12 Some key results obtained in the Indo-UK Center on Glass-Ceramics for Osteoporosis (**a**) Research Highlights (**b**) Micro-computed Tomography Analysis: Bone volume (BV) to total volume (TV) ratio for glass-ceramic implant (LG26Sr) ($4.5SiO_2$-$3Al_2O_3$-$1.5P_2O_5$-$3SrO$-$2SrF_2$) and commercial implant (HA-bioglass; HABG), in rabbit femur for different time periods. (*/#): Significant difference (statistical) in terms of BV/TV ratio in case of HA-bioglass implant with respect to 4 weeks implantation data at $p < 0.05$. The data are represented as mean $\pm$ SE (Sabareeswaran et al. 2013)

program, the Indo-UK Biomaterials centre on glass ceramics for osteoporosis, at IIT Kanpur (http://www.iitk.ac.in/UKIERI_biomaterials). The collaborating institutes were SCTIMST, India, and three UK Universities (Birmingham, Warwick, and Kent). Altogether, six co-PIs and 15 researchers have taken part in this interdisciplinary research consortium. The implantation experiments with Sr-doped glass ceramics in a rabbit animal model confirmed neobone formation, qualitatively using histological analysis and quantitatively using micro-computed tomography (Fig. 3.12). The osteoconductive property is comparable to that of a commercially available bioactive implant (HA-based bioglass). This project has led to the development of novel glass-ceramic implants that have been found to be biocompatible *in vivo*, with regard to local effects after implantation, besides training a number of young researchers from India and the UK.

3.10 Nota Bene

Several research groups at IISc Bangalore, IIT Kanpur, IIT Guwahati, IIT Kharagpur, IIT Bombay, VNIT Nagpur, NIT Rourkela, IIT Delhi, SCTIMST, CSIR-CGCRI, CSIR-NML, CSIR-CLRI, IIT Kharagpur, PSGIAS, IIT Roorkee, IIT (BHU) Varanasi, IIEST Shibpur, CIPET Chennai, IIT Hyderabad, and SIB (AIIMS) have been actively working on biomedical implants for orthopaedic, dental restoration, cardiovascular, ophthalmic, ENT, cartilage and neurological applications, tissue and regenerative engineering, and 3D bioprinting of biomaterials, tissues, and organs. This has led to two unique centres of excellence—BETiC lead by Prof. B. Ravi at IIT Bombay and the other is at IISc, Bangalore, sponsored by the DBT. BETiC, (a consortium of IIT Bombay, Tata Memorial Centre and VNIT Nagpur), is one of the success stories. Recently, IIT Kanpur has initiated a centre of excellence in collaboration with King George's Medical University, Lucknow. Several micro, small, and medium-scale enterprises (MSMEs) are now working to indigenise various medical implants and devices that have been developed over the last several years.

A number of challenges, however, restrict more intense translational research when taking laboratory-scale research to market. These challenges are presented in Chap. 5 of this monograph. Many of these challenges are significantly addressed in North America, Europe, and Australia. These facilitated significant translational research in these continents, as discussed in Chap. 4 of this monograph.

3.11 A Path Ahead

The last three decades have shown that there is tremendous zeal and zest in the community of biomaterials scientists to develop a new generation of biomaterials, medical devices and implants, while providing newer insights into the science of biocompatibility. I hope that key changes, that are suggested in this monograph, will usher in an era of new and significant achievements, building upon the knowledge base, enthusiasm, and hard work of the scientists, clinicians, entrepreneurs, administrators, and science and technology facilitators of India.

Chapter 4
International Status

Abstract The quantum of translational research in developed countries is generally high, with a better research ecosystem, having hospitals attached to many research centres. As we have seen in the previous chapter, high quality research is ongoing in academia and national labs, in India. However, the ecosystem in India is not yet sufficiently developed to allow a streamlined and efficient approach to product development and translation. In many developed nations, there are well-developed regulatory frameworks in place, better connectivity between biomedical manufacturers and private companies, more corporate funding for research, and well-developed contract research organisations. This chapter is intended to highlight the translational research activities, which are key examples of successful research ecosystems. The examples shown in the geographical map are no way meant to be an exhaustive list, but represent the work of major multi-institutional research centres around the globe. An overview of the work of major research groups is summarised in Appendix C.

Disclaimer: The presentation of material and details in maps used in this chapter does not imply the expression of any opinion whatsoever on the part of the Publisher or Author concerning the legal status of any country, area or territory or of its authorities, or concerning the delimitation of its borders. The depiction and use of boundaries, geographic names and related data shown on maps and included in lists, tables, documents, and databases in this chapter are not warranted to be error free nor do they necessarily imply official endorsement or acceptance by the Publisher or Author.

B. Basu, *Biomaterials Science and Implants*,
https://doi.org/10.1007/978-981-15-6918-0_4

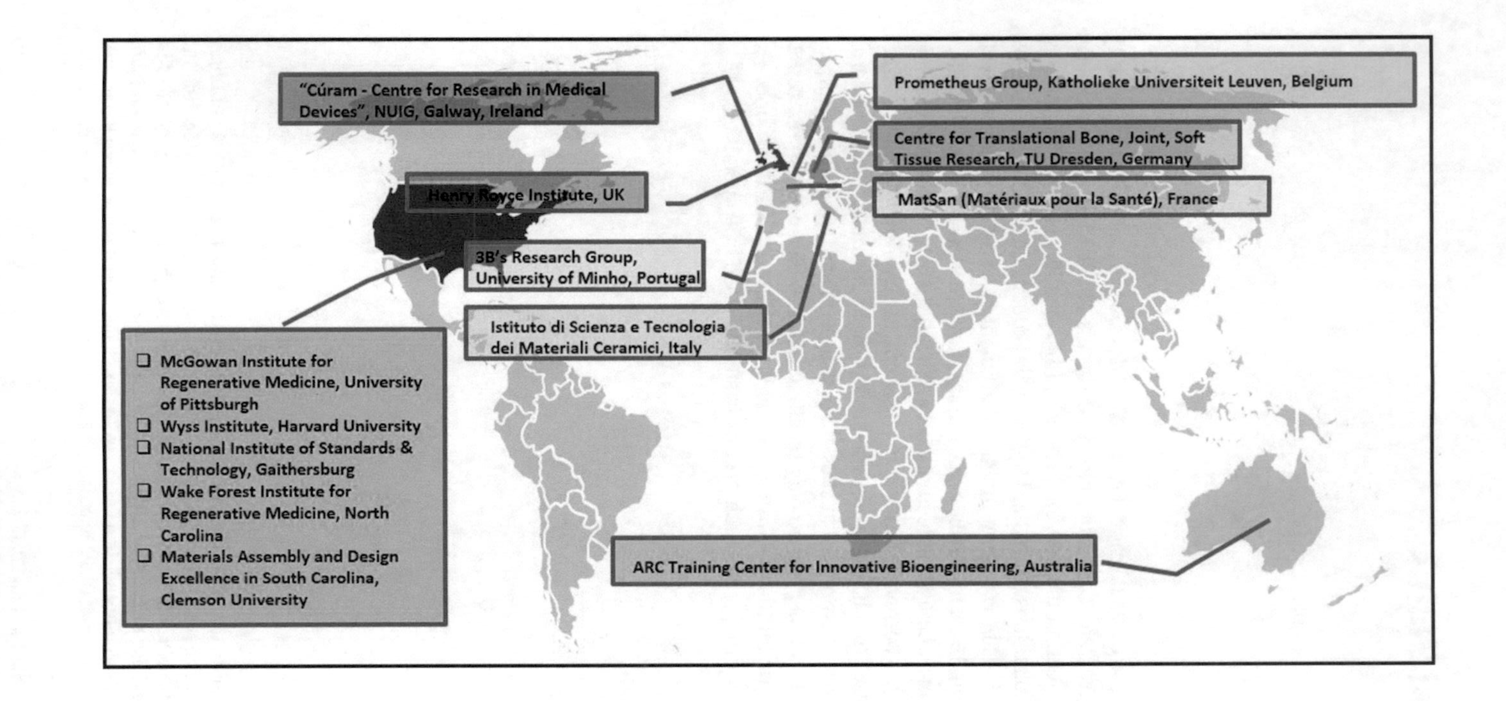
"Cúram - Centre for Research in Medical Devices", NUIG, Galway, Ireland
Henry Royce Institute, UK
3B's Research Group, University of Minho, Portugal
Istituto di Scienza e Tecnologia dei Materiali Ceramici, Italy
McGowan Institute for Regenerative Medicine, University of Pittsburgh
Wyss Institute, Harvard University
National Institute of Standards & Technology, Gaithersburg
Wake Forest Institute for Regenerative Medicine, North Carolina
Materials Assembly and Design Excellence in South Carolina, Clemson University
Prometheus Group, Katholieke Universiteit Leuven, Belgium
Centre for Translational Bone, Joint, Soft Tissue Research, TU Dresden, Germany
MatSan (Matériaux pour la Santé), France
ARC Training Center for Innovative Bioengineering, Australia

4.1 Global Landscape of Biomaterials Research Consortia

4.1.1 Prometheus Group, at Katholieke Universiteit Leuven, Belgium

The Prometheus Group of Katholieke Universiteit Leuven, Belgium, is a translational platform in the field of regenerative medicine.

This is one of the finest examples of a group of clinician scientists and bioengineers working within the ecosystem of a university hospital.

The group aims to bring tissue-engineered bone grafts from the laboratory to clinical practice, by including scaling up, reproducibility, quantification, quality control, surgical training, and economic feasibility as a part of the translational programmes. The group's areas of focus are fundamental science, translation of scientific findings into manufacturing processes, clinical translation, and development of enabling technologies, such as computer modelling, imaging, bioreactors, and sensing. In the first phase, clinically relevant *in vitro* and *in vivo* screening tools for the quantitative evaluation of potential bone growth-stimulating compounds (cells, signalling molecules, materials/coatings, and drugs) are being developed. In the second phase, these tools will be used to produce and to validate bone tissue engineering concepts.

4.1.2 McGowan Institute for Regenerative Medicine, University of Pittsburgh, USA

The McGowan Institute is a multidisciplinary network of over 250 faculty members whose primary academic appointments are in 32 different academic departments of the University of Pittsburgh. The institute has multiple programmes, which are aligned in four pillars, namely, tissue engineering, cell-based therapies, medical devices and artificial organs, and clinical translation. There is a strong commitment to ensuring that the technologies developed by McGowan-affiliated faculty are commercially/clinically available.

To date, there are 31 start-up companies that are the result of technologies developed by McGowan-affiliated faculty.

The key areas, that are being explored, include biomaterials science, and organ engineering—either via 3D printing, or decellularised organs.

4.1.3 'Cúram - Centre for Research in Medical Devices', NUIG Galway, Ireland

The Cúram centre is a national centre, which hosts collaborations across 61 academic leads, 500 researchers, 8 clinicians, and 6 universities. The clinicians are the key opinion leaders and have led medical device trials. Cúram collaborates with 28 industry partners, 10 multinational corporations and 18 small-to-medium scale enterprises. The centre has 406 academic collaborations globally. The centre's vision is to develop affordable, innovative, and transformative device-based solutions to treat global chronic diseases (see Fig. 4.1). The clinical areas of interest at the Cúram Centre are respiratory, neural, cardiovascular, musculoskeletal, renal and urology, and soft tissue. These areas have been selected based on the presence of academic expertise in the area, relevant industries in Ireland and key opinion leaders in the area.

The Cúram Centre is based on the premise that industry is central to drive the translation of research into the next generation of medical devices and implants.

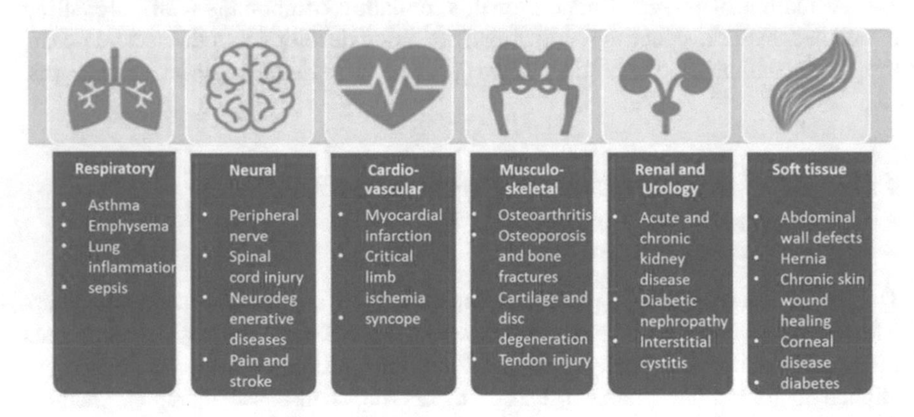

Fig. 4.1 Overview of research areas of the Cúram Centre

4.1.4 *Wyss Institute at Harvard University, USA*

Harvard University's Wyss Institute is a signature example of a coexisting community of scientists, clinicians, entrepreneurs, and business executives.

> The institute is focused towards promoting biomedical products through stages up to commercialisation, based on a model for innovation, collaboration, and technology translation.

The Wyss Institute focuses on developing new bioinspired materials and devices for applications in healthcare among other areas. The Wyss Institute currently has 18 core faculty members and 15 associate faculty members from Harvard and its partner institutions, who are leaders in the field of biologically inspired engineering.

Technical scientists and engineers from the advanced technology team bring industrial experience in product development and team management to the Wyss Institute. The team helps to build and lead integrated technology development teams focused on high-value applications. The 'machine shop' hosts full-time engineers and houses state-of-the-art manufacturing equipment that enables fast prototyping and construction of devices. The institute's business development team as well as experienced entrepreneurs-in-residence (EIRs) facilitate the translation of the technologies from the laboratory to commercialisation. The team also assists with intellectual property, technology licensing, and/or start-up company opportunities, and develops collaborations and strategic partnerships with industry, clinicians, and entrepreneurs (see Fig. 4.2). Strong collaborations with industry partners enable the understanding of markets and user needs to match the breadth of applications, envisioned by institute researchers.

4.1.5 *Centre for Translational Bone, Joint, and Soft Tissue Research, TU Dresden, Germany*

The Centre for Translational Bone, Joint, and Soft Tissue Research was established in 2010 to facilitate collaborations among researchers from University Hospital Dresden and Medical Faculty Carl Gustav Carus of Technische Universität Dresden, Germany. This centre has a focus to strengthen the experimental research of orthopaedics, trauma, and reconstructive surgery, and oral and maxillofacial surgery. Around 40 scientists, PhD students, and technicians are working in this centre. The researchers of this centre work closely with industry and international researchers in translational research, e.g., 3D bioprinting of tissues in collaboration with Gesim GmbH. At the centre, the research on 3D bioprinting is focusing on extrusion-based 3D printing (3D

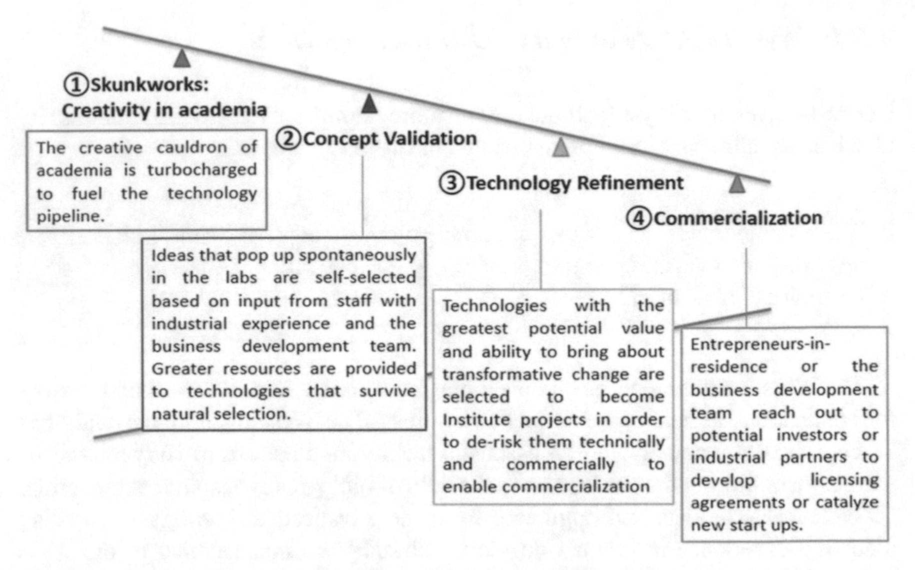

Fig. 4.2 Translational ecosystem within Harvard University's Wyss Institute

plotting) and a 3D bioprinting approach, using micro-algae for medical and biotechnological applications. The researchers are also working on therapeutic drug/protein delivery biomimetic bone matrices and biomaterials for articular cartilage defect healing.

Bioprinting for space exploratory missions using strontium-substituted calcium phosphate bone cements is currently being pursued.

4.1.6 Istituto di Scienza e Tecnologia dei Materiali Ceramici (ISTEC), Faenza, Italy

In the field of regenerative medicine, there are several ongoing programmes at ISTEC, Italy. A number of researchers at ISTEC are working on the development of biomimetic materials (3D structures) from bioceramics and hybrid composites (see Fig. 4.3). The 3D structures are recognised and recruited in the human body by the natural metabolic processes. Further, they instruct cells to trigger and sustain tissue regeneration. The biomaterials are developed by novel nature-inspired fabrication approaches to generate innovative materials capable of smart performance, such as inherent tissue-inductive and antimicrobial properties with unprecedented mechanical performance. The research on new functionalities, triggering remote activation

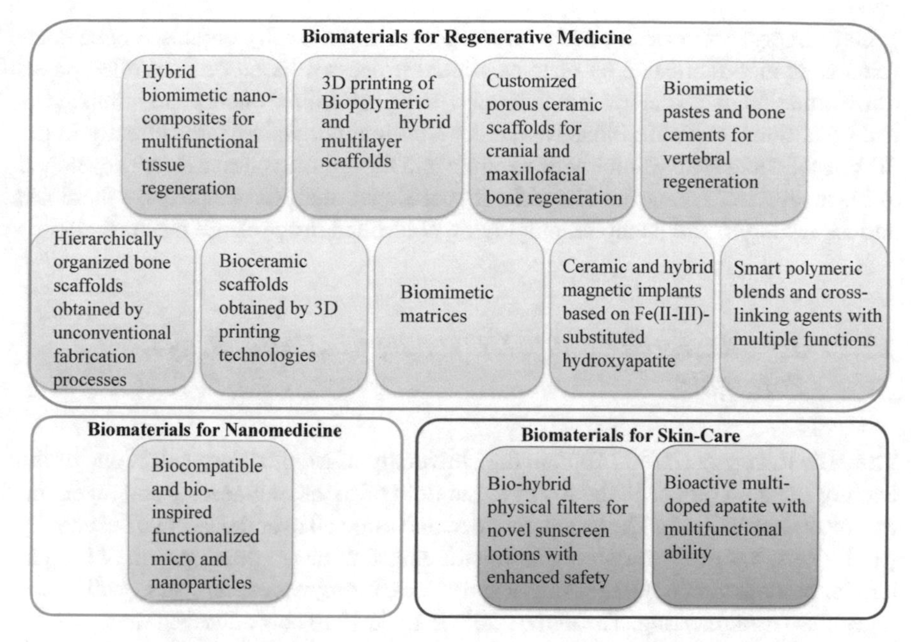

Fig. 4.3 Research areas of Istituto di Scienza e Tecnologia dei Materiali Ceramici (ISTEC), Italy

and/or enabling *in vivo* guiding, is a current topic of high interest. This approach, which involves the recruitment of endogenous biologic factors, paves the way towards personalised medicine.

Hybrid biomimetic nanocomposites are being developed for multifunctional tissue regeneration. These are obtained by bioinspired mineralisation of natural polymers with mineral biomimetic nanophases. The scaffold can be obtained with graded mineralisation and porosity, to mimic the complex tissue structure of osteochondral and periodontal regions.

Recently, 3D printing of micro-extrudable ceramic-based, biopolymeric, and hybrid inks are adopted to develop 3D engineered (stroma-like) organomorphic scaffolds mimicking the complex architecture of specific bony regions, soft tissues, and organs.

At ISTEC, customised porous ceramic scaffolds for cranial and maxillofacial bone regeneration were developed. These were obtained by direct foaming of calcium phosphate suspensions with rheology optimised to yield final constructs with interconnected macroporosity and maximised mechanical strength. Bioactive pastes and bone cements have also been developed as injectable formulations based on ion-doped calcium phosphates and natural polymers, targeted to bone regeneration in spinal surgery, orthopaedics and neurosurgery. Hierarchically organised bone scaffolds were also fabricated by unconventional processes based on controlled phase transformation of 3D natural templates occurring at minimal energy and in supercritical conditions, to obtain bioactive nanocrystalline inorganic phases directly in the 3D consolidated state. Biomimetic ceramic and bio-hybrid matrices were developed as 3D models of the bone marrow niche for predictive analysis of anticancer therapies and for testing *in situ* therapies, employing controlled drug release mechanisms.

4.1.7 3B's Research Group, University of Minho, Braga, Portugal

The 3B's Research Group (3B's) in the University of Minho, Portugal, is one of the leading research groups in the world in the field of tissue engineering and regenerative medicine (TERM). The group, founded and directed over the last two decades by Rui L. Reis, has made long-term, landmark contributions to development of biomaterials, biodegradable, natural origin polymers, biomimetics, and stem cell-based regenerative engineering. The aim of 3B's is to develop new advanced therapies for the regeneration of human tissues, and the team comprises an international team of more than 200 researchers.

> Since its foundation, the 3B's group is recognised for the development of novel biomaterials (hydrogels, scaffolds, membranes, nano/micro-particles, smart/functional surfaces) from natural polymers, including marine origin polymers, for applications in drug delivery and tissue engineering of bone, cartilage, skin, intervertebral disc, tendon, meniscus, and neurological tissues.

3B's has founded four different spin-offs companies. The research infrastructure facilities are specifically designed to execute state-of-the-art tissue engineering-related research, such as new materials processing, stem cells isolation and differentiation, and *in vitro* and *in vivo* biocompatibility testing.

4.1.8 National Institute of Standards & Technology (NIST), Gaithersburg, MD, USA

The Alliance for Regenerative Medicine conducted a survey of pharma companies and found that 'product consistency and lack of standards is possibly the single greatest challenge being faced in the field' of regenerative medicine. Similarly, the Regenerative Medicine Foundation conducted a stakeholder survey which concluded that the lack of reference materials to benchmark measurements and validation criteria for critical assays is the most significant hurdle being faced by the industry.

Several standard guides exist that address scaffolds from a generalised perspective. However, due to the intense interest in fibre-based constructs, there is a need for a guide that focuses specifically on fibres. The guide will provide a compendium of techniques for characterising fibre-based constructs and guidance on selecting appropriate tests based on the application.

Fibre-based constructs include constructs made by electrospinning, forcespinning, meltspinning, pneumatospinning, extrusion, drawing, and other related processes. A workshop was held to discuss the guide and to serve as a launching point for the working group, which currently consists of 33 stakeholders, with the majority being from industry. A second working group has been formed to develop a standard test method for measuring cell viability in a scaffold. Tissue-engineered medical products often consist of cells suspended in a 3D scaffold, and the product's mechanism of action may rely on the cells being viable to regenerate tissue. However, the scaffold hinders cell viability measurements in a variety of ways. Thus, a working group has been formed to develop a model scaffold-cell-assay system for assessing cell viability in scaffolds. An interlaboratory study will be used to assess the reproducibility of the method in different laboratories. Reproducible measurement systems are the key during biomanufacturing when thousands of units need to be assessed for consistency. These results will provide a strategy for validating measurements of cell viability in scaffolds and will support the development of the standard guide.

> To address clinically unmet needs, the researchers in the Biomaterials Group at NIST are leading the development of a new documentary standard guide for characterising fibre-based constructs for tissue-engineered medical products.

4.1.9 Wake Forest Institute for Regenerative Medicine, Winston-Salem, North Carolina, USA

The Wake Forest Institute for Regenerative Medicine (WFIRM), USA, is recognised as an international leader in translating scientific discovery into clinical therapies, with many world firsts, including the development and implantation of the first engineered organ in a patient. More than 450 scientists collaborate on regenerative medicine research at this institute, the largest in the world. The major achievements of institute scientists include engineering replacement tissues and organs in four categories: flat structures, tubular tissues, hollow organs, and solid organs.

WFIRM's scientists have successfully implanted several tissues and organs such as skin, urethras, cartilage, bladders, muscle, and vaginal organs, which have been created in the laboratory using a variety of engineering strategies. To scale up the manufacturing process, WFIRM turned to 3D printing and over many years, the institute's scientists have developed the integrated tissue and organ printing system. They demonstrated that it is feasible to print human scale living tissue structures that have the right size, strength, and function for use in humans. The scientists have optimised the water-based 'ink' that holds the cells so that it promotes cell health and growth. They also have printed a lattice of micro-channels throughout the structures to allow nutrients and oxygen from the body to diffuse into the structures and keep them alive, while they develop a system of blood vessels.

The institute's expertise in generating micro-human organs and tissues from the cellular level to functional tissues led to its selection to lead a $20 million 'Body-on-a-Chip' programme. The goal was to build a miniaturised system of multi-human organs to model the body's response to chemical and biological agents and develop potential therapies. The Body-on-a-Chip platform can accelerate the translation of basic scientific discoveries into the clinic, in a faster and more accurate way than in laboratory animals.

The WFIRM team was the first to create miniature tissue equivalents using regenerative medicine techniques that replicate a high-level, long-term function, and they were able to combine up to six organs on the same device.

4.2 Multi-institutional Research Initiatives and Training Centres

4.2.1 Henry Royce Institute, UK

The Henry Royce Institute is a national centre bringing together 800 researchers for collaborating in the broad area of advanced materials in the United Kingdom. This institute is founded at the University of Manchester and its research hub is spread across the founding partners that are the Universities of Cambridge, Liverpool, Leeds, Oxford, Imperial College London, the National Nuclear Laboratory and the Culham Centre for Fusion Energy.

The Royce Institute has nine critical research themes, one of which is biomedical materials. The capability landscape is shown in Fig. 4.4. The aim of the biomedical materials area is to provide direct patient benefit and to improve medical testing and device production. An example of an application of biomedical materials would be to create 3D *in vitro* tissues that improve pharmaceutical testing and also reduce the need for animal testing. Key research outcomes or targets include bespoke hard and soft implants created by additive manufacturing, new multifunctional fibrous scaffolds for tissue regeneration, new bioelectronic systems, faster biomedical device testing facility, minimally invasive tissue repair, and smart materials for remote sensing or monitoring (e-health).

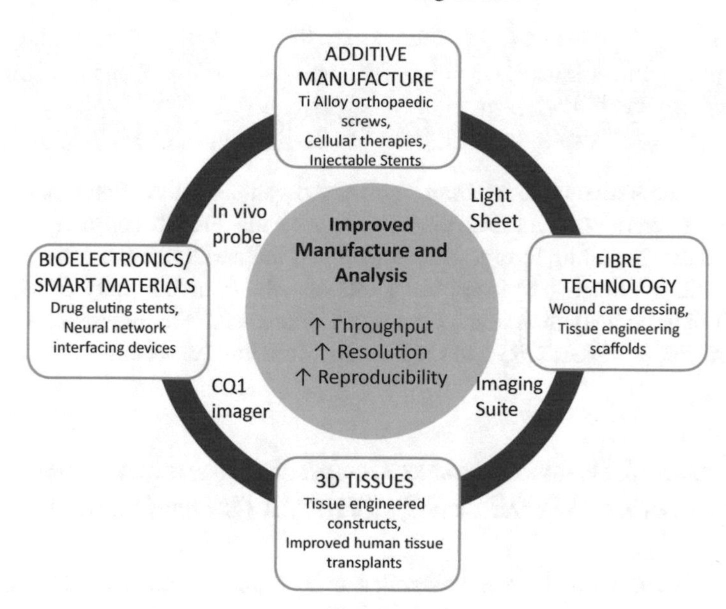

Fig. 4.4 Capability landscape of the Henry Royce Institute (adapted from original figure from the Henry Royce Institute)

The focus of the Royce's Biomedical Materials area is Fibre Technology, Bioelectronics Medicine, Additive Manufacturing and product scaling up to TRL 5 Level.

4.2.2 Matériaux pour la Santé (MatSan), France

MatSan is the French 'Materials for Health' commission. MatSan gathers together more than 65 academic and over ten industrial centres, as well as clinicians. The research themes include bioceramics/bioactive glasses and mineral/organic composites for bone regeneration, (bio)polymers and hydrogels, stimuli-responsive polymers for (soft) tissue repair, metals and alloys for bone implant applications, cancer treatment, medical diagnosis, additive manufacturing, coatings, injectable pastes, biomimetism, the study of biomineralisations, and exploration of biological tissues.

Special attention is paid to the synthesis of biomaterials and their precursors, which may be metastable, to their physical–chemical characterisation, their processing (dense or porous scaffolds, films and coatings, injectable pastes, additive manufacturing/3D printing, colloidal particles, etc.), and the property evaluation (durability/ageing, biocompatibility, etc.).

On a collaborative note, MatSan regularly organises international and national conferences, thereby increasing the visibility of the French community towards other partners, including foreign researchers and industrial representatives. MatSan includes four national-scale scientific societies—the French society of Metallurgy and Materials (SF2M), the French Group of Ceramics (GFC), the French Center for anti-corrosion (CEFRACOR) and the French "Titanium" association.

4.2.3 Materials Assembly and Design Excellence in South Carolina (MADE in SC), Clemson University, USA

Materials Assembly and Design Excellence in South Carolina (MADE in SC) is a new five-year venture, established in September 2017, with funding from the National Science Foundation. The overarching vision of the programme is to create a global materials innovation hub as a statewide effort with ten participating institutions in South Carolina. Nearly, over 65 faculty members are engaged in training and mentoring of research scholars (40 undergraduate and 30 graduate per year) and post doctoral researchers. Additionally, it also provides training and mentoring of STEM high school teachers, partnerships and collaborations among team members and external partners, and shared materials research infrastructure. The focus is on multi- and inter-disciplinary research with three major thrust areas, which are unified by an integrated modelling and computational core (Fig. 4.5).

Thrust 1 is on hierarchical organic and inorganic complexes of materials, primarily for enhanced optical, electrochemical, and magnetic properties. Thrust 2 is on soft materials, mainly polymers. The emphasis is on designing stimuli-responsive polymers, that can interact and respond to the changes in their environment. For example, pH-responsive polymers and hydrogels for wound healing is one of the focal areas. Thrust 3 is on rational design of biomaterials. The focus areas of this thrust are synthesis and biogenic assembly, integration of biomaterials into micro-engineered devices and biological test beds.

A second area of interest at MADE in SC involves the design of micro-engineered platforms to investigate the responses of cells to topology, mechanical forces, and electric fields and development of 3D constructs from extruded biomaterials.

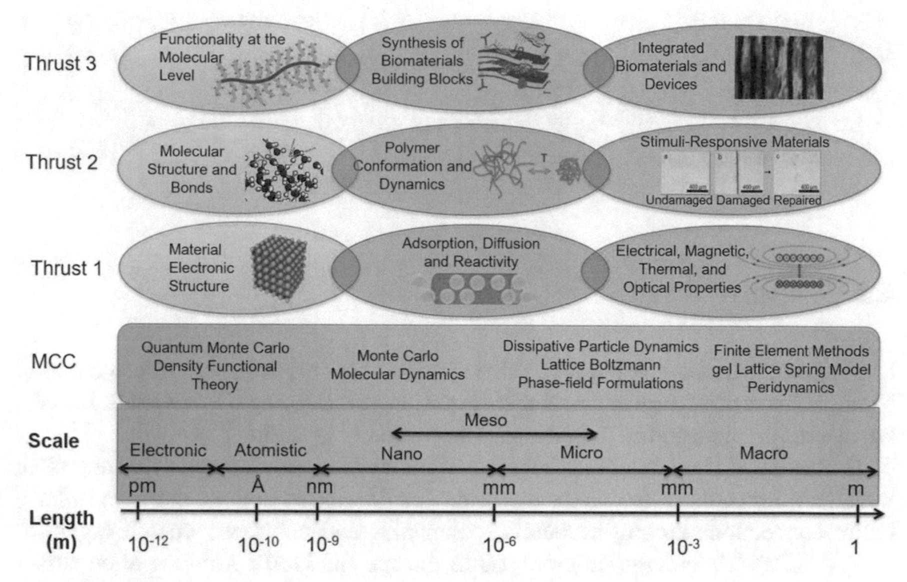

Fig. 4.5 Overview of the scientific areas of the MADE in SC initiative (Image courtesy Rajendra K. Bordia)

4.2.4 ARC Training Centre for Innovative Bioengineering, University of Sydney, Australia

The Australian Research Council (ARC) training centre for innovative bioengineering is devoted to address the uprising need of personalised medicine for the diverse age groups in the growing population of Australia. ARC training centre's goal is to bring together early career researchers, industries, and hospitals to innovate new technology followed by translating those towards 'new treatments'. The centre is spearheaded by threefold goals, 'advance training', 'develop and support students', and 'help students find their place'. At present, the ARC-sponsored centre functions in close collaboration with seven universities and five industry partners.

There are three prevalent research themes in the ARC training centre. The theme 'technologies', where the primary research interest is focused mainly on 'pioneering diagnostic therapeutic wearable and implanted technologies' is accompanied by the second research theme of 'innovative biomaterials'. The development of new advanced biomaterials, design, and fabrication is one of the basic research interests in this division. 'Sensors and telemetry technologies' is the third category of the threefold research theme, where the 'continuous' evaluation and optimisation of future implants are being performed. In terms of training, there are currently two 'training themes' available in ARC, namely TT1 and TT2.

In TT1, the centre provides hands on training in translating new technology into an industry setting, while TT2 provides frameworks for developing routes to market for innovative biomedical technology solutions.

4.3 Closure

In developed nations, a better research ecosystem not only enhances research output but also allows proximity of researchers to patients undergoing clinical trials. In Asia, we can see an opportunity to better the innovation ecosystem.

Examples of well-developed regulatory frameworks are CE and FDA approval, which are followed in the western world. The developed nations are seen to have better connections among academia, companies and contract research organisations (CROs). Many large programmes in Europe and North America often attract significant funding from private companies.

Transforming India's innovation ecosystem requires due thought and careful planning. In order to achieve a more robust and resilient research ecosystem, sustained efforts are needed to emulate the best practices and policies of well developed ecosystems, while adapting these to the Indian context. For this purpose, I present an overview of the challenges involved, and also key recommendations for the path ahead, in Chaps. 5 and 6, respectively. A key part of improving the research ecosystem is improving the education system which supplies the human resources. This will be discussed in further detail in Chap. 6. I envision that together, all stakeholders can make concerted efforts towards ensuring that the Indian research ecosystem attracts and retains the best talent, and harnesses its collaborative synergies towards addressing the challenge of providing high quality affordable healthcare. Unmet needs have historically driven many great scientific discoveries. Now is the time for us to discover not only such discoveries, but also the full potential of a robust research ecosystem that can become a role model for many developing countries. It is hoped that the next two chapters will encourage greater dialogue on this subject, and that the actionable suggestions presented will be of help in this respect.

Chapter 5
A Challenging Frontier and Status of National Policies

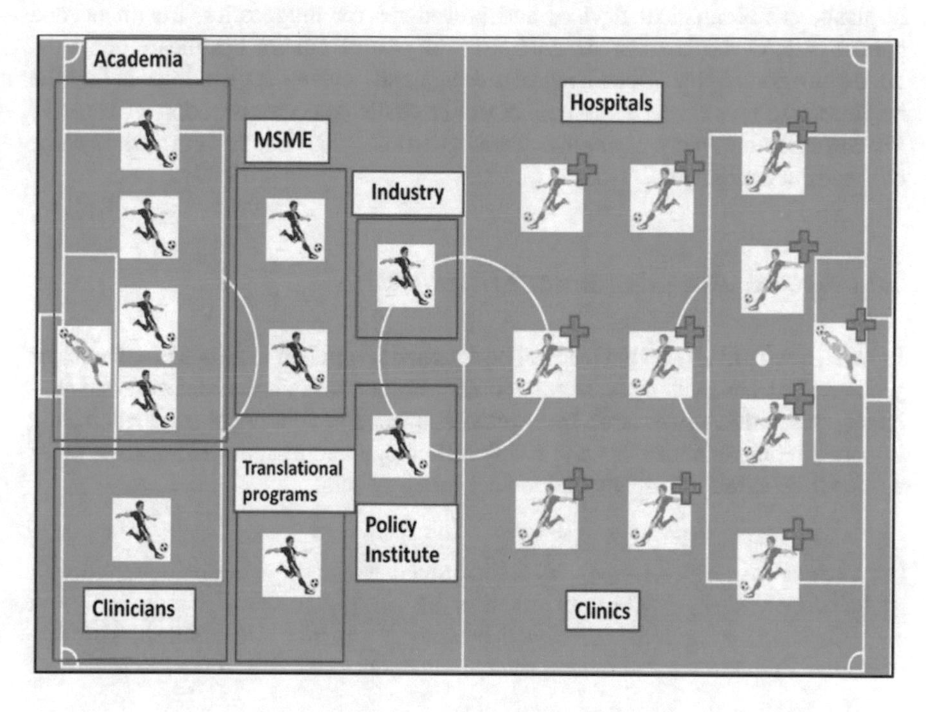

Abstract It is hard enough to mimic, predict, and create the intricate and elegant structures of living tissue or organs and harder still to make them sustainable and affordable for the common man. Over the years, the field of biomaterials science has blossomed with impressive strides in establishing connections with nanoscale science, genetics, and molecular and cell biology. Along with the coming together of multidisciplinary teams of physicists, chemists, biologists and engineers; new opportunities for understanding, manipulating, and mimicking biological materials and processes have opened up. With unique materials and devices, biomaterials have emerged as a frontier science, challenging the very foundations of accepted paradigms and pushing the boundaries of existing knowledge, especially around its interface with human health.

B. Basu, *Biomaterials Science and Implants*,
https://doi.org/10.1007/978-981-15-6918-0_5

As discussed in Chap. 2, the global markets for biomaterial implants in different segments like orthopaedic and dental implant/device segments are expected to increase steadily in the coming decade. Yet, the march of science is fraught with challenges. Although in the past few decades, the focus on patient-specific solutions has risen extensively, there are worrying questions that refuse to go away: How can scientists tap into the enormous unaddressed potential of biosystems and biomaterialomics? How far can clinical and societal needs be best met? What biological strategies would lead to materials that build and repair themselves? How can biomaterials be made multifunctional, dynamic, responsive and sustainable?

A few developing nations like India and China have seen a significant improvement in the quality and efficacy of healthcare due to gradual acceptance of imported implants and biomedical devices and procedures for diagnostics, treatment, and patient care, in the last few decades. But the cost of quality healthcare has escalated tremendously; how can scientists design and develop applications so that the healthcare solutions reach the economically challenged cross-section of society? While addressing many such unanswered questions, this chapter summarises some of the major challenges.

5.1 Accelerating Regulatory Approval

In India, access to affordable medical devices and healthcare facilities is still a major problem. Current challenges are collaboration among biomaterial scientists and clinicians, translation of research from prototype stage to functional products, costly imported biomedical devices and viable technology and manufacturing scale up of academic research outputs by biomedical entrepreneurs.

A strong regulatory framework is of immediate importance in India, since Indian companies need to obtain CE or US FDA approvals to penetrate into global markets. In India, there is a need for enforcement of standards to ensure that product quality and patient safety remain uncompromised.

5.1.1 Current Status and Gaps in Medical Device Certification

To improve access to safe, accessible, affordable, and appropriate medical devices, a number of policies and related teams have been implemented since 2014. These include the Medical Device reform task force (2014/15), Inverted Duty Correction for spared/components (2016), Medical Device Rules (2017), preferential procurement of domestically manufactured goods, GFR (2017), Medical Devices R&D

priority and market access think tank- KIHT (2017), Code for Marketing Practices (2017), Medical Devices Promotion Council (2017), and Medi Valley (2018). Also, the government is thinking of creating a separate Ministry of Pharmaceuticals and Medical Devices to serve the sector in a more integrated manner.

The Medical Devices Rules (2017) stipulated the rules for manufacturing, import and sale of medical devices, as well as for clinical trials, labelling, and inspection.

Effective from 2018, it is expected that Medical Device Rules will bring the regulatory system in India to be on par with global standards. Medical devices will be classified on the basis of risk: high-risk devices are to be regulated by the Drug Controller General of India (DCGI) and low-to-moderate risk devices are to be regulated by the appropriate state authority. The devices should conform to specific standards applicable, e.g. ISO. It is also indicated that the shelf life of the devices should not exceed 60 months, unless satisfactory evidence is provided by the manufacturer to justify an extension.

Regarding quality certification for medical devices in India, the Association of Indian Medical Device Industry (AIMED), in collaboration with the Quality Council of India (QCI) and the National Accreditation Board for Certification Bodies (NABCB), is implementing a voluntary quality certification scheme for medical devices, known as the Indian Certification for Medical Devices (ICMED) Scheme. The scheme is expected to instill confidence among buyers, while diminishing the credibility of substandard products.

In the above context, KIHT's Non-regulatory Innovation Potential Utility and Novelty Certificate (NIPUN) is the first ever indigenous testing and certification programme for medtech innovations to access the Indian market. The NIPUN scheme provides help in business plan analysis, quality compliance, testing, rapid prototyping, and market access.

DBT-BIRAC recently initiated the National BioPharma Mission (including Medical Devices), whereby industry-academia consultation paved way for government funding in the health technology sector.

The South East Asia Regulatory Network (SEARN) is responsible for regulation of medical products, including medicines, vaccines, biological, and medical devices and diagnostics for human use in the eleven countries of the South East Asia Region.

The Central Drugs Standard Control Organization (CDSCO) is the national regulatory body for Indian pharmaceuticals and medical devices. To advise CDSCO on matters related to the regulation and implementation of medical devices, the Union Health Ministry has constituted a special committee, the Medical Devices Technical

Advisory Group (MDTAG) in 2019. Besides the Chairperson and the Member Secretary, 22 other members constitute the committee from various departments containing representatives from organisations, including Department of Science and Technology (DST), National Institute of Biologicals (NIB) and Indian Council of Medical Research (ICMR) and Association of Indian Medical Device Industry (AIMED), Sree Chitra Tirunal Institute for Medical Sciences and Technology (SCTIMST), IIT Delhi, Association of Diagnostics Manufacturers of India (ADMI), Bureau of Indian Standards (BIS), Medical Technology Association of India (MTAI), Atomic Energy Regulatory Board (AERB), Defence Research and Development Organization (DRDO), Department of Electronics and Information Technology (DEITY), National Health Systems Resource Centre (NHSRC), Federation of Indian Chambers and Commerce Industry (FICCI), Confederation of Indian Industry (CII), and American Chamber of Commerce in India (AMCHAM).

The mandate of MDTAG is to discuss issues relating to the implementation of Medical Devices Regulations, while aiming to strengthen the regulations to make medical device companies accountable for quality and safety of their products.

Further, they will take up matters with the Drugs Consultative Committee (DCC) and Drugs Technical Advisory Board (DTAB).

5.1.2 Manufacturing-Related Regulations

It is mandatory that the manufacturers of the biomedical implants and devices follow ISO 13485:2016, an international standard to ensure the compliance with Good Manufacturing Practices (GMP), which governs the medical devices quality management systems for regulatory purposes. The standard had undergone a major overhaul in 2016. The emphasis of the regulators used to be more on design, design risk and design validation and complaint management, with less emphasis on manufacturing. However, a shift happened to place more emphasis on analysing the manufacturing risks (PFMEA) and identifying the methods to prevent problems from happening (risk management and control). The standard governs all activities of the product life cycle, from conception to disposal.

Three key aspects of the ISO 13485:2016 standard are validation, evidence, and contamination control.

The aim of validation is to certify that steps in the routine manufacturing process will not go wrong. Validation requires submission of a process map, process risks, installation qualifications for equipment, critical process parameters that influence outcome, definition of worst-case parts with rationale, acceptance test methods, process challenge at extremes of process limits, and process limits set as the control limits for routine production. The particular requirements for implantable medical devices are that the organisation must provide records of components, materials, and conditions for the work environment used. Typically, precision machining is used for biomedical implants.

Most implants include few critical features with tolerances of less than or equal to 0.05 mm.

Validating machining process is optional, if 100% verification is practiced. Notably, many companies choose to validate the machining process to reduce the inspection burden. For cleaning and passivation, validation is mandatory. This is an area, where the respective regulatory professionals are going through advanced training. For proving cleanliness, one has to prove that there are no contaminants from the manufacturing process and also from the cleaning process. Residues should be negligible or within safe limits. Also, the manufacturing process should not adversely affect biocompatibility.

Accelerated ageing, drop testing, and sterility testing are some of the methods used to test packaging, so that reliable data can be provided to justify the indicated shelf life. Regarding sterilisation, typical methods include gamma irradiation, ethylene oxide gas, and gas plasma (H_2O_2). The selection of a specific sterilisation protocol depends on the type/composition of the implant. Validation of sterility assurance level is mandatory.

5.2 Strengthening Basic Research in Biomaterials Science

The human body has limited ability to induce self-repair and regeneration. Various tissues of the musculoskeletal, nervous, cardiovascular and urological systems rarely achieve functional regeneration, post-injury. Multifunctional biomaterials with tissue-specific biocompatibility are clearly needed.

One of the unique challenges therefore lies in identifying new biomaterials beyond the limited range, currently licensed for use in humans.

In the context of the evolution of biomaterials, the first generation of biomaterials was used to achieve a combination of physical properties to match those of the tissue being replaced, with minimal host response. Examples include high-strength materials such as titanium and steel, as bone implants. While developing second-generation biomaterials, importance was given to materials that could elicit a desired response in a controlled physiological environment, while simultaneously minimising the immune response. There has been an increasing need to develop biomaterials with multifunctional properties that can facilitate a more natural way for repair and regeneration of host tissues.

Third-generation biomaterials are designed to elicit specific reactions at cellular and subcellular levels.

5.2.1 Development of Next-Generation Biomaterials

The biggest challenge lies in developing high-quality indigenous biomedical implants, in a field that is growing faster due to technological advancement and societal needs. Nearly 6 million people die each year as a result of injuries, a global public health problem. And the number of people at very advanced age, globally, is surging, projected to triple between 2017 and 2050. Implants cater to over 2 million such people every year, whose quality of life gets compromised by debilitating injuries. They need medical implants of excellent biocompatibility, superior corrosion resistance, and high strength-to-weight ratio. Also, the number of revision surgeries has increased at about the same rate as that of primary surgeries, largely due to prosthesis failures. More than 30% of the failures are directly due to post-surgical infection (Fig. 5.1). To avoid this, on-site delivery of antibiotics during post-surgery period is one of the clinical options. A summary of the ethical and safety considerations of biomaterials research is provided in Appendix B.

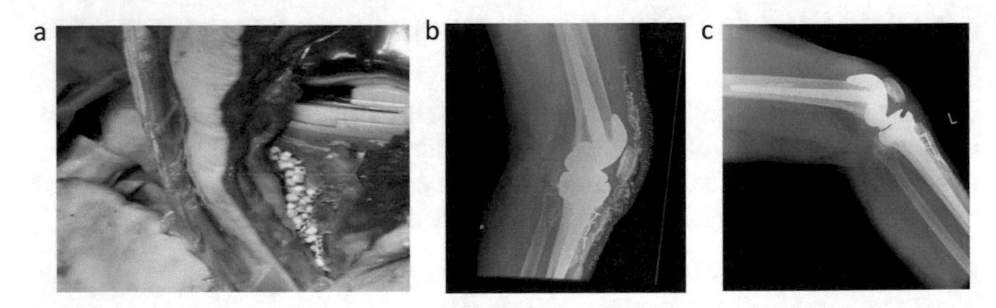

Fig. 5.1 Infected total knee replacement second-stage operation: Initial treatment carried out with long-term antibiotics due to medical risk assessment. Two-stage revision planned due to progressive loosening. (**a**) Intra-operative X-ray clinical image, (**b**) Post-operative X-ray image to reveal soft tissue, (**c**) Six month post-operative X-ray to show full absorption and infection-free state

This clinical challenge can be alternatively addressed by modifying metallic implants to release drugs, post-implantation, without affecting their structural, mechanical and biological attributes.

It has been anticipated that by 2030, the demand for primary THR will increase by 171% for patients with less than 65 years of age. In 2015, among all joint replacement procedures, total knee replacement (TKR) showed the highest market contribution. The number of TKR surgeries is expected to increase by 4 million cases by 2030, exclusively in USA. In spite of overwhelming accomplishments, the natural joint-like functioning of UHMWPE has not yet been achieved in many applications (viz. hip, knee, shoulder, elbow, ankle and spine), due to performance-limiting clinically relevant properties, especially aseptic loosening due to wear.

A better clinical outcome in many musculoskeletal surgeries demands the development of UHMWPE-based hybrid composites, either blends with another biocompatible polymer, or with carbonaceous reinforcements to enhance wear resistance property.

With respect to implants, the selection of biomedical grade raw materials is important in producing desired products. For bioceramic implants, grinding and polishing are critical challenges for production, and appropriate machinery needs to be established for THR implants. Furthermore, addressing the stiffness mismatch between the implant and surrounding bone is of major concern since it can result in osteolysis, or active resorption of the bone by osteoclasts (Fig. 5.2).

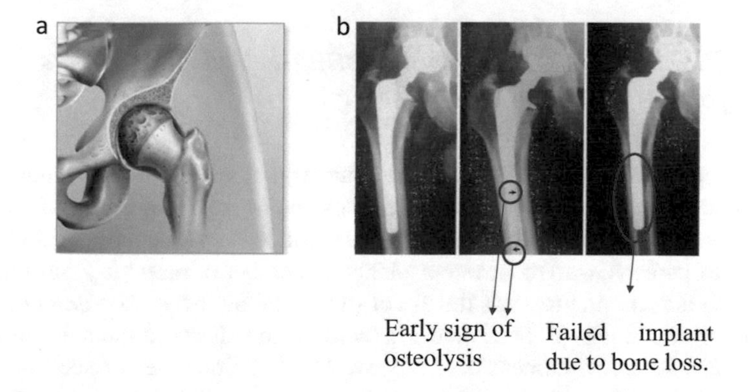

Fig. 5.2 (**a**) Diseased hip joint, (**b**) Stiffness mismatch and failure of implants (post-operation radiography images)

In the case of the musculoskeletal implants, the subject-specificity and bone condition are to be critically considered, while designing the next-generation implants.

In order to address such challenges, a new generation of acetabular liners or dental implants are to be developed and clinically tested. Another area in orthopaedics, where biomaterials play an important role, is the use of calcium phosphates as bone filler in orthopaedic/dental surgery or as bone cement or for drug delivery applications. An important challenge here is the prediction and validation of sustained drug release from nanomaterial scaffold-based delivery systems for treatment of life-threatening diseases, including cancer. Researchers from pharmaceutical sciences can play a major role in this, from the development of drug-eluting coronary stents, drug- releasing orthopaedic devices to tissue engineered scaffolds for targeted drug delivery applications.

Several Indian research groups are currently working on doped hydroxyapatite (HA), tricalcium phosphate (TCP), and biphasic calcium phosphate (BCP) for orthopaedic and dental applications. Doped HA-based materials have been developed for drug delivery and hyperthermia treatment in cancer therapy, too.

While several competing wet chemical synthesis processes using synthetic chemicals or biogenic minerals are being investigated, it is important to develop commercially scalable synthesis techniques to develop products—powders, granules, and porous blocks—made of materials like hydroxyapatite for various medical applications.

5.2.2 Biomaterials and 3D Bioprinting-Based Approaches for Regenerative Engineering

Due to accidents or diseases, humans often undergo traumatic injuries in their lives which lead to loss of function in many tissues and organs. However, the supply of donor organs is limited. It becomes an arduous task to supply healthy organs to large numbers of patients, who require transplants. A number of research groups in India and abroad are working towards the use of functional biomaterials-based platforms, and further combining of these platforms with stem cell-based therapies to mimic the natural micro-environment of the body. The functionalities of such platforms provide 3D matrices with desired porosities, elasticity, and wettability along with topographical cues that favour cellular attachment, growth, and differentiation, as well as proper nutrient flow for tissue regeneration. With respect to cartilage tissue engineering, cell implantation within the articular joint poses several problems. Also, optimisation of scaffolds for cartilage tissue is needed, together with a good fit of the scaffold to the perimeter of the defect. Maintaining the desired hyaline cartilage phenotype is also another challenge.

Although, several impressive tissue engineering approaches have been reported in the field of regenerative engineering over the past decades, successful transfer of these results into clinical trial evaluation still remains a challenge.

Biofabrication is an emerging field, wherein complex biologically functional tissues are fabricated using extracellular matrix, growth factors, living cells and biomaterials. The aim is to fabricate native-like tissue structure or organs, with specific dimensions and high reproducibility by 3D printing. 3D bioprinting techniques are the most recently developed biofabrication approaches in which pore geometry, distribution, and interconnectivity with precise deposition of cells within the patient specific scaffold geometry can be fabricated to improve cell signaling, cell-cell and cell-ECM interactions, cellular viability and functionality within the bioprinted construct. Bio-inks are the cell-laden biomaterial that can be integrated into the additive manufacturing process to mimic the extracellular matrix environment and support the living cells so that cell adhesion, proliferation, and differentiation take place within the 3D printed construct.

To investigate the modulation of innate immune response, macrophage polarisation pathways can be explored in tailoring the vascularisation of scaffolds.

At several IITs, different research groups (S. Ghosh in IIT Delhi, R. Banerjee in IIT Bombay and S. Dhara in IIT Kharagpur) are working on synthetic and natural polymers for tissue engineering and regenerative engineering applications. In this context, the modulation of viscoclastic property is one of the essential pre-requisites to regulate the cell functionality for regenerative medicine. The use of microrheology techniques, like DWS needs to be used in this context.

In evaluating the biocompatibility, the researchers mostly use conventional petri dish culture of single cell type (osteoblasts, fibroblasts, chondrocytes, Schwann cells, neuroblastoma, etc.), *in vitro*. It is important to understand cellular changes at the gene-level, and how the cells sense cues in the cellular micro-environment on a biomaterial platform. Biomaterials researchers use several biochemical assays for cytocompatibility without focusing much on cell signaling pathways. The experiments involving co-culture under physiologically relevant conditions are also not widely conducted in the field of biomaterials science. While significant efforts are underway to develop tissue engineered scaffolds for organ bioengineering applications, the immunogenicity of synthetic biomaterials, eliciting innate or adaptive immune response, is not widely explored.

With respect to the neurosurgical application, peripheral nerve injuries (PNIs) are a very common type of nerve injury, and restoration of such kind of nerve damage remains a greater challenge in the field of regenerative engineering (Fig. 5.3). The conventional clinical treatment has many limitations and disadvantages, such as improper fitting and loss of native functionality, donor site morbidity, pre-existing injury to the donor nerve, donor-host mismatch and host immune response.

For neurosurgical treatments, the development of nerve conduits of gaps larger than 10 mm remains a challenge.

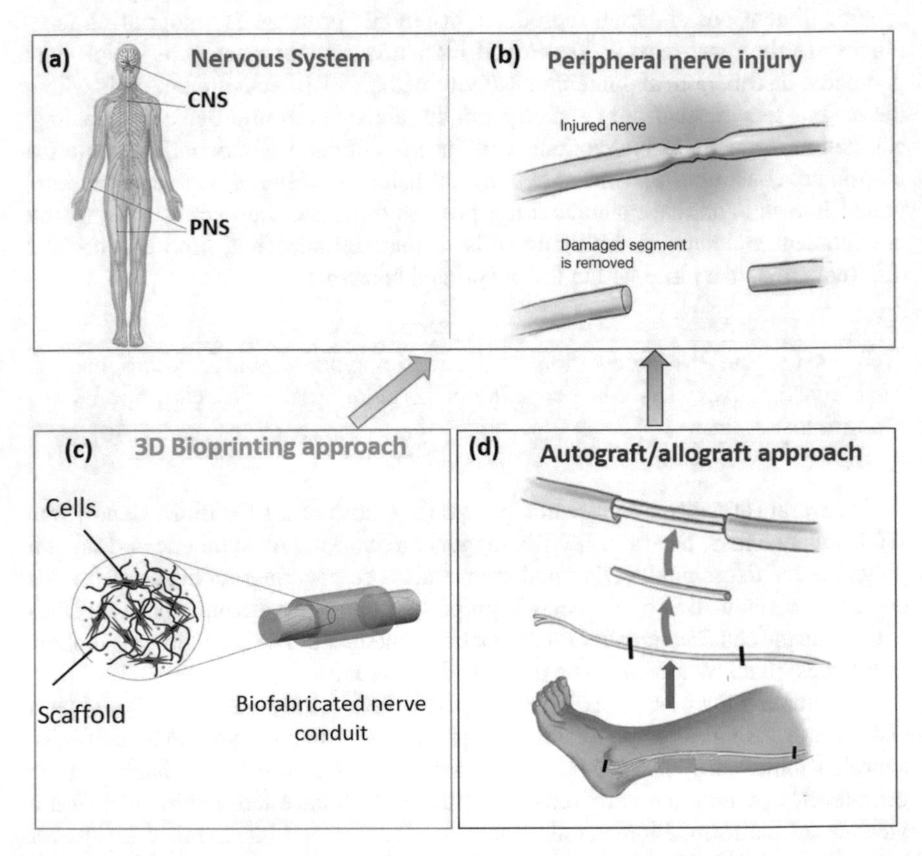

Fig. 5.3 (**a**) Overview of the anatomy of the central nervous system (CNS) and peripheral nervous system (PNS). (**b**) Peripheral nerve injury (*Source* Mayo Foundation). (**c**) 3D bioprinting approach for biofabrication of a nerve conduit for peripheral nerve repair (Hu et al. 2016). (**d**) Nerve graft approach for peripheral nerve repair (*Source* Mayo Foundation)

Therefore, future research aims should be directed to develop an artificial nerve conduit for the substitution of injured peripheral nerves with patient-specific architecture, optimum mechanical properties, conductivity, and biodegradability.

In the field of urological reconstruction, defects of the human urinary bladder and urethra result from cancer, tumour, loss of elasticity, insufficient capacity (for bladder), and detrusor compliance. These are common and serious problems, specifically in the elderly patients.

The repair and restoration of human urothelial tissues are challenging problems due to the extremely weak regenerative tendency of urothelial cell lines.

Concerning the urological pathologies, bladder carcinoma (BC) generally originates in urothelium. Subsequently, the tumour invades the detrusor muscle layer, leading to what is clinically known as muscle invasive bladder cancer (MIBC). As per the available statistics, around 25–30% of all the bladder cancers is MIBC, which necessitates surgical intervention to partly or completely remove the bladder tissue, in an effort to restore the normal urological physiology. In addition, neurological conditions and congenital bladder anomalies result in a functionally deficient urinary system.

The typical procedure for end stage bladder disease involves replacement or augmentation with a section of the patient's gastrointestinal tract as an autologous graft. Despite significant efforts to develop biomaterials for urinary bladder reconstruction, autograft of the ileum still remains as 'gold standard', due to the lack of acceptable clinical performance of synthetic products.

The primary challenge in the development of biomaterials for urological reconstructive applications is to use two different cell lines in different hierarchical tissue layers, since the inner surface of urological tissues contain the urothelial cell lines and the smooth muscle tissues are present in the outer surface. Therefore, future bioprinting strategies must fabricate bilayered scaffolds for bladder reconstruction or a cylindrical urethra. To achieve appropriate tissue extensibility compared to native bladder tissue, selection of fabrication approach and choice of materials are important.

5.2.3 Multiscale Computational Analysis

The use of calcium phosphate in clinical applications was highlighted earlier in this chapter. For (multi)-ion doped calcium phosphate bioceramics, the local chemical environment surrounding the dopant in HA/TCP-lattice structure governs the dissolution of ions, which significantly influences the biomineralisation and cell functionality. It is important to pursue ultrastructural and computational analyses to correlate combinatorial research involving micro-structure with functional properties and biocompatibility to allow better control over structure–property correlation in bone-mimicking materials.

The recent results from several Indian research groups suggest that the modulation of stem cell fate processes can be achieved by stimulating them with endogenous bioelectric fields. However, several key questions remain to be answered. These questions are related to the specific mechanisms of stem cell activation by external electrical stimuli.

Significant lessons are to be learnt by correlating the electric field parameters and level of tissue or substrate conductivity required to trigger differentiation of choice, if these variants function in harmony to govern stem cell behaviour.

Through the research of IISc/IIT (BHU) researchers, piezoelectric and conducting materials, in addition to external stimulation (mechanical and electrical), form the basis of new, interesting and efficient strategies for bone tissue engineering and regenerative medicine applications. Owing to the number of advantages associated with piezoelectric biomaterials (piezoelectricity, pyroelectricity and ferroelectricity), and biological (poling-induced mineralisation and cellular) response, it is envisaged that piezoelectric biomaterials could serve as orthopaedic implant materials in the near future. IISc/IIT (BHU) researchers have developed electric/magnetic field stimulation protocols to modulate cell functionality on several biomaterials.

The fundamental principles that drive the synergistic interactions among substrate properties and stimulation parameters to govern cell-material interactions are still unknown.

This is also an area, which can be benefitted by multi-scale computational analysis. Also, more pre-clinical studies conducted on piezobiomaterials in larger animal models, followed by clinical trials in human subjects are needed.

The interaction between biomolecules (proteins) and the biomaterial surface needs to be probed at a single molecular level, using computational tools, like, MD/Monte Carlo simulation. This is the key step governing the cell-material interaction.

Despite the recent focus at the interface of biology and engineering in academia, India lacks extensive coordinated translational research programs on biomaterials and implants.

For this, there are several bottlenecks that need to be dealt with. A major challenge in the field is that it is largely based on experimental findings, based on multi-scale and multidisciplinary measurements—electron microscopy, micro-computed tomography for 3D micro-structure analysis, cell viability and differentiation assays, and pre-clinical and clinical outcome assessment. Less effort, however, has gone into multi-scale models: molecular dynamics, micro-mechanical finite element models and so on.

5.2.4 Data Science Driven Approach

Despite several decades of research, we still do not have proven capability to predict how to design biomaterials that are capable of interacting specifically with the biological system of a patient. The concept of biomaterialomics would allow us to analyse and integrate existing computational tools with the knowledge of biomaterial 'genes' and accelerate the process of biomedical device innovation. The 'genes' of biomaterials would include the structure and properties. In line with the existing concepts of 'proteomics', 'genomics', and 'materials genome', the alternative phrases to describe the above concept can be 'biomaterialome', 'biomateriome', or 'biomaterials genome'.

After an extensive discussion among the thought leaders in the area of biomaterials science at Chengdu, China, in June 2018, the term 'biomaterialomics' was conceived and defined as the 'integration of computational tools, large databases and experimental techniques, to explore the basic material elements and combine them to discover and design new biomaterials for medical products.'

This concept would be based on the application of the theory and technology of materials genomics to biomaterials.

The conventional approaches in biomaterials science or bioengineering involve the intuitive way of tailoring process variables that often incur long development cycles and high costs (Fig. 5.4). Against the backdrop of the ever-increasing unmet clinical needs, it is therefore important to develop patient-customised, implantable biomaterials or biomedical devices in an accelerated manner to accomplish the bedside-bench-bedside translation cycle. Building on the concepts of the Materials Genome Initiative, it is recommended to adopt 'biomaterialomics'. This new concept should rely on a data-driven, integrated understanding of biocompatibility, and elements of biomaterials development. The newer computational approaches to establish process-structure-property (PSP) linkages, established in the field of materials science for structural materials, can play significant roles in developing the biomaterialomics concept.

It should be emphasised here that the biomaterialomics approach can significantly aid in the development of new-generation bioimplants, whose predictive clinical performance would be closely tracked by 'digital twins'. An example is provided in Fig. 5.5, which also presents a new concept, as how computational data science based approaches can be merged with experimental and multiscale physical modeling approaches in near future.

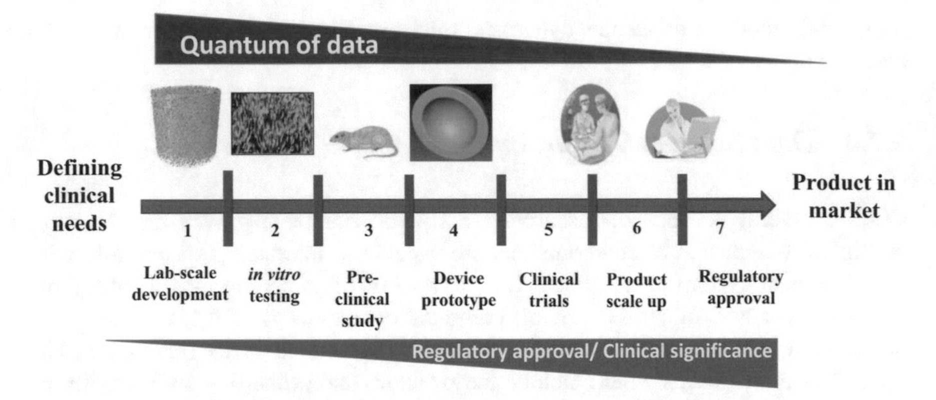

Fig. 5.4 Typical innovation workflow in biomaterials/biomedical device development, showing the qualitative trend in quantum of data and clinical significance at various phases

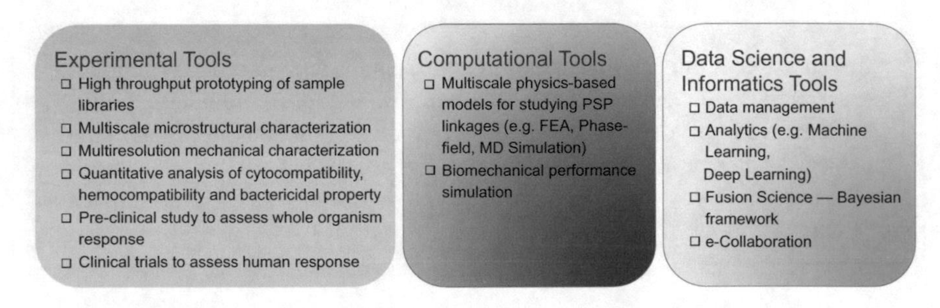

Fig. 5.5 Summary of the major elements used to build a biomaterialomics framework

To substantiate further, Fig. 5.6 presents the concept of the machine learning (ML) algorithm, which can be adapted to accelerate the discovery of new biomaterials. The quantum of high-quality data required in the ML approach unfortunately requires significant experimental efforts, including many control experiments.

The Biomaterialomics approach will bring together concepts and ideas from artificial intelligence (AI), machine learning, deep learning, etc., and will combine with the results of the computational approaches (e.g. FEA-based biomechanical analysis or MD simulation results) employed in the field of biomaterials science.

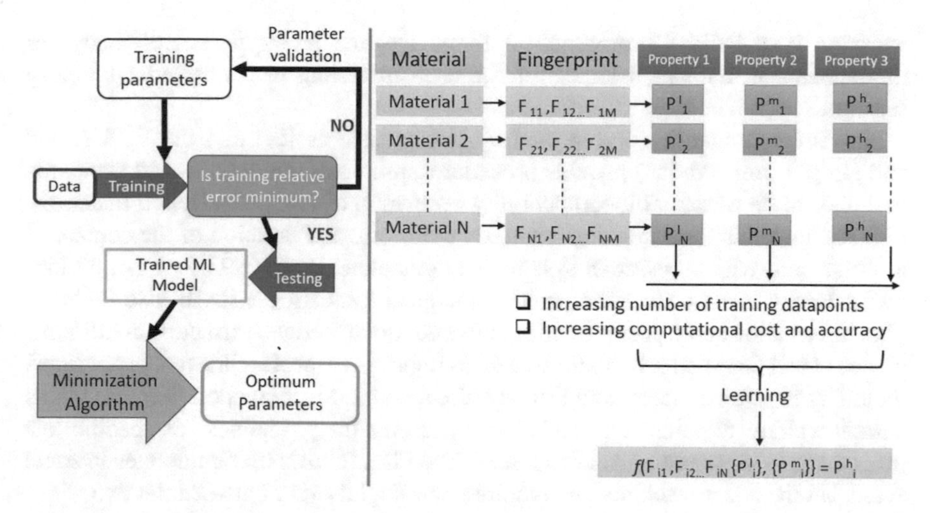

Fig. 5.6 Overview of the machine learning approach, which can be employed in the field of biomaterials science

5.3 Building Translational Research for Human Healthcare

Translational research converts 'bench' basic research discoveries into products that can be used at the 'bedside' of a patient in a hospital clinic, as part of healthcare treatment and prevention. Translational research involves researchers, clinicians, start-ups, industries and policy makers.

The domain of translational research aims to improve the accessibility of healthcare by accelerating scientific discoveries related to biomedical implants and biomaterials from the institutional laboratories to clinical trials and then, the transfer of the technology to start-ups and industries.

Here, the institute network in the Indian context can comprise of academia, start-up companies, constituent laboratories of Council of Scientific & Industrial Research (CSIR), Department of Science and Technology (DST) and Department of Biotechnology (DBT), DRDO/DAE laboratories and regulatory consultants.

The very first stage of this translational research is the identification of the unmet clinical need and the human disease model.

Once the problem is recognised, it is taken up by the institute networks, as mentioned above. The project teams in the institute network, which includes faculty members, project researchers, graduate students and consultants or entrepreneurs, work on it to build a prototype. At this stage, the prototype is not ready for implantation in humans, because of the need of testing in controlled laboratory settings.

Government agencies and regulatory bodies such as ISO and the FDA (Food and Drug Administration) provide procedures, protocols, guidelines, and standards that need to be used for biocompatibility evaluation of newly developed materials. ISO recommends standards depending on the nature and duration of the contact of a biomaterial with an osseous system. The guideline, ISO 10993-Part 1, provides methodology for choosing the proper biological evaluation tests. It also includes vital information about positive and negative control materials, extraction conditions, choice of cell lines and cell media, as well as important aspects of the text procedures, including tests on extracts, and tests by direct and indirect contact. Part 2 explains animal welfare requirements, and Part 3 presents the guidelines for specific test procedures or other testing-related issues. The FDA follows ISO guidelines in some areas, but the test procedures and requirements slightly vary between the two.

All non-clinical laboratory studies, including *in vitro* and *in vivo* studies on biomedical materials and implants, should comply with the Good Laboratory Practice (GLP) regulations.

Understanding of the regulatory compliance and documentation of the research in the exploratory phase as per the industry requirements, generally assists in licensing the technology to established companies more easily. The commercialisation of medical devices involves appealing, not just to individual consumers but to a complicated landscape of stakeholders, from doctors and patients to regulators and insurers, all of whom have a say in whether a new technology is adopted. At the Sree Chitra Tirunal Institute of Medical Science and Technology, Thiruvananthapuram, a cardiac surgeon has led a large team of researchers in developing artificial heart valves a few decades ago, which were further commercialised in the Indian market by the TTK Healthcare group. Since then, some attempts are being made for commercialisation of products, but not to a significant extent.

For the translation of biomedical research, innovative technologies need to be taken from academic institutes to the market and then to hospitals and clinics for affordable healthcare. For this, the technology first needs to be developed and evaluated by academia in collaboration with clinicians. Then, Indian patent and patents under Patent Cooperation Treaty (PCT), can be filed at host institutes and this patented technology would be transferred to the start-ups or micro, small, and mediumscale enterprises (MSME). The technology, when matured to a higher Technology Readiness Level (TRL), can be licensed to the industries (Fig. 5.7). With the help of industries and policy institutes, it is possible to market the products to hospitals and clinics.

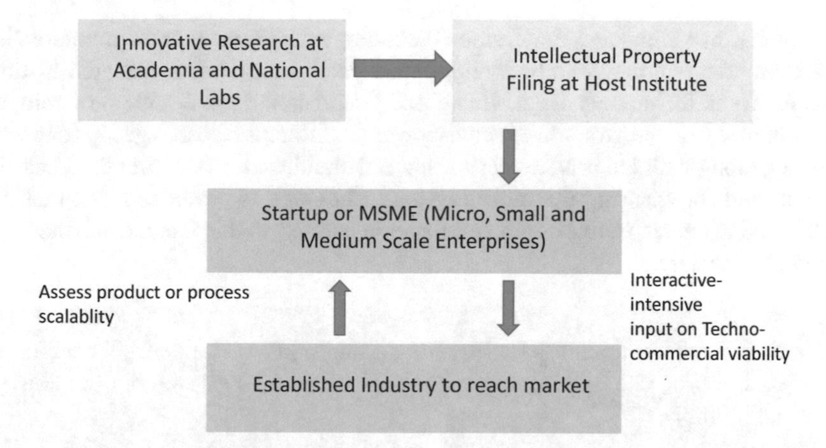

Fig. 5.7 Translation ecosystem model, showing the pathways for technology transfer from the innovation hub

> To bring biomedical technologies to market, the team of academia, clinicians, MSMEs and Industry, has to play their individual role in a very coordinated manner, to bring the product to market.

New products undergo several considerations for market entry; including product scalability, breadth of application area and cost effectiveness. Typically, after the first prototype is developed and tested, a few size variants of the product undergo more extensive testing. Some products, like the total knee joint replacement (TKR) system or dental implants require an armamentarium of components, tools and devices. Sometimes instruments for facilitation of accurate positioning of implants during surgery are also required. The translational research team therefore needs to accelerate the design, make, test, characterise and iterate cycle (the entire product development process), and couple this to advances in machine learning.

> We need an 'Internet of biomaterials' to advance translational research from laboratory to the market faster.

5.4 Nourishing Industry Collaborations, Incubators, and Start-Ups

In the healthcare system, the challenge is to create an ecosystem that brings innovations into education, society, hospitals, services, and governance, and one that is directed towards transforming the lives of people, the poor, and rural sectors in particular. It is not sufficient to develop a technology or product, but it should be commercially viable and deployed for treatment/diagnosis of the relevant population across the country, in a stipulated timeframe. Technology deployment and commercialisation is an area that needs to be strengthened. Each translational research institute should have a Technology Deployment and Commercialization Cell for deploying the technology on the ground. It is envisaged that different research groups can work together, along with the private sectors and stakeholders, for pursuing the identified projects and for ensuring that not only the technology is developed, but that it is transferred to the respective commercialisation partner, so that it can reach the target segment of society.

In the context of implant-based treatment, high cost or limited affordability by common people has been a major concern in many developing nations, including India.

More so, as more than 85% of the fragmented market is dominated by imported implants. Currently, around 1 million patients need prostheses and implants every year, with India importing Rs 7000 crores worth of them per year. Indigenously developed implants are likely to reduce the total treatment cost, and are thus a step towards providing affordable healthcare.

Commercially, a robust understanding of design elements as well as commercial requirements aids researchers in overcoming the challenges of bringing a new biomaterial to the market.

Currently, industry often does not recruit sufficiently skilled personnel for developing biomedical devices for local needs. A great number of talented graduates are employed by the medical device groups—TCS, WIPRO, Cognizant and Infosys—for servicing their clients in the USA or Europe. An excessive emphasis on low cost also leads to the inability of industry to pay enough to attract and to retain the right talent. It is hardly a surprise that the domestic orthopaedic industry, has very few PhD or postgraduate engineers, despite having an annual turnover of Rs 1500 crores.

As evolution of a technology occurs, ideally, it will attract more resources for clinical testing, manufacturing, and commercialisation. There is a fine balance between accruing adequate clinical data and getting the product into the market at the earliest. In this case, it is essential to follow internationally accepted norms. Based on these, one can decide on the threshold number of trial surgeries, adverse event frequency considerations and follow-up time of patients. Alongside, market evolution occurs, sometimes leading to new application areas and new treatment paradigms. These, in turn, affect product development.

Another challenge is market pushback. This can result in lowering the product's cost, better packaging, increased projection of the presence in the market and marketing in other countries.

Given these changes, there is a need for industry experts to be involved at the time of commercialisation.

In joining translational research consortia, there are challenges faced by MSMEs. Firstly, the response time is slow, and many MSMEs do not want to take the risk of penetrating the market alone. They prefer established industry and larger players and facilitators like KIHT to be involved in helping them reach the market and in assisting them with business plans. Also, in some cases, the biomedical area is not the only area of focus, and they lack experience in commercialising products in this area, particularly in the Indian context.

To achieve significant progress, industry should be aware that the process of idea translation involves the crossing of three 'valleys of death', where the idea, prototype and product have to undergo various challenges and risks.

From idea, to functional prototype and then to the market—there are gaps known as 'valleys of death'. Crossing them depends on the product type, process scalability, economics, commercial competitiveness and many other factors (Fig. 5.8). It is interesting to notice that while a deep valley can be expected at the translational stage, such a valley is shallow but wider at the commercial stage.

Some of the emerging challenges faced by the Indian biomaterials community include inadequate linkages among academia, research and industry, ageing of the institutional framework, weak innovation ecosystem to convert ideas into the research programs/useful products and processes. Poor coupling among several key aspects needs to be addressed. They are technology, trade, and social development needs, low base of full-time equivalent scientists per million population and science and

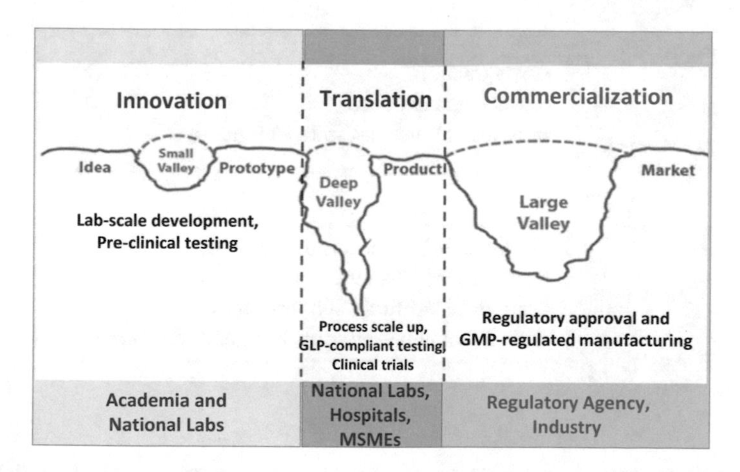

Fig. 5.8 Schematic illustration of the 'valleys of death' involved in the translational research on biomaterials and implants

technology manpower base. A balance between basic research and well-directed and focused research has to be maintained. India should have allocated manpower to enhance the interfacing among clinicians, industry and academia for translation of healthcare products and for fostering these linkages and facilitating technology transfers. Such staff would help to drive the translational projects and to assist in procuring funding for projects. Also, more start-ups supported by venture capital would be able to introduce new products at competitive prices and to create new job opportunities in the Indian context. The sequence of chronological growth stages for a start-up or an incubatee is shown in Fig. 5.9.

While a start-up may have technologically smart founders, and it is relatively easy to take the invention from idea to validation stage, more difficulties would be faced as one progreses towards the scale up stage.

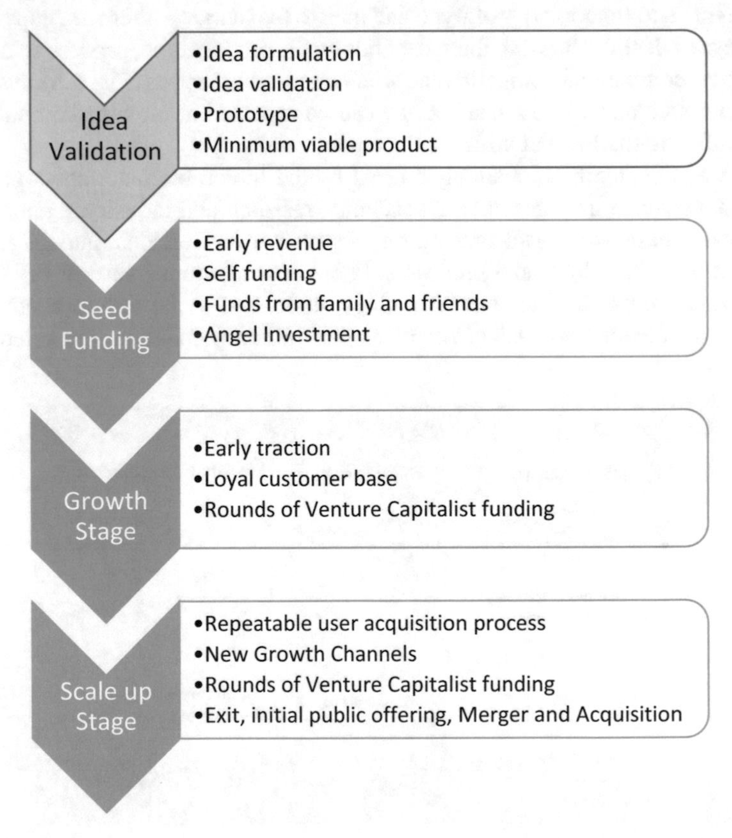

Fig. 5.9 The various chronological stages of start-up growth (*Source* National Report on Government of India's States' startup ranking 2018—www.startupindia.gov.in)

5.5 Addressing the Gaps in Medical Education and the Clinical Research Ecosystem

5.5.1 Education and Training

To accelerate indigenous implant development, the involvement of clinician scientists should be emphasised. Pre-clinically studied biocompatible implants or devices need to undergo clinical trials in human patients. Apart from the human ethics committee approval, clinical studies require strict regulatory approval and funding to support them. Once a device is proven to provide clinically appropriate patient outcomes, the technology for device manufacturing can be transferred to biomedical device companies. The involvement of clinicians at the ideation stage and the clinical trial stage is therefore extremely significant.

The translational ecosystem in India significantly lacks the involvement of clinician researchers and this is a challenge of the first order.

In India, intermittent efforts were invested to streamline and to support the medical education system. The guidelines or recommendations of the various expert committees were included in the report on Medical Education and Support Manpower (1975), Indian Council for Social Science Research – Indian Council for Medical Research (ICSSR–ICMR) (1981), National Health Policy (1983), the Medical Education Review Committee (1983), National Policy on Education (1986), the Expert Committee on Health Manpower Planning, Production and Management (1987) and more recently, Niti Aayog (2015).

The current medical curriculum places little emphasis on research or bioengineering-focused projects, both at the undergraduate and postgraduate levels.

Currently, Phase I of the medical curriculum includes preclinical subjects of anatomy, human physiology and biochemistry (15 months). Phase II covers pathology, microbiology, pharmacology, forensic medicine and community medicine (18 months). Phase III covers medicine and surgery, including orthopaedics, obstetrics and gynaecology, paediatrics and eye and ENT medicine. Clinical postings begin in Phase II, and Phase III includes a community medical posting. After passing the MBBS, candidates are granted provisional registration for one year, during which they undergo a compulsory rotating internship, which includes hospital training and community health work.

Also, science-based subjects, like cell and molecular biology, including microscopy-based analysis, biostatistics and fundamentals of biomedical engineering must be taught in the undergraduate medical curriculum.

An immediate aim should be to build a vibrant education and research ecosystem in existing AIIMS and other national medical institutions. To achieve this, a bioengineering immersion programme (internship programme) should be offered to undergraduate medicine students and clinical immersion program should be offered to the graduate students in bioengineering.

An MD/PhD program needs to be introduced for clinicians and also important would be an external registration PhD program for biodesign innovators in academic institutions.

5.5.2 *Clinical Research Ecosystem Development*

The existing ecosystem is lacking in opportunities for clinicians to pursue research work, and medical colleges are regularly promoting faculty without a track record of mentoring researchers or doing research. Promotions require that faculty should have at least one or a few publications, but the Medical Council of India (MCI) does not specify the level of authorship required. The medical system is significantly affected by the time that is given to government practice versus private practice. Often, clinicians are not very well paid at government institutions, the facilities are not adequate and the patient load is very high.

In such a situation, it is understandable as to why the motivation for clinicians to spend time on research is low, and this is a challenge for addressing unmet clinical needs through research innovation in medical institutions. Also, the number of research collaborations and the number of multicentric clinical trials are on the lower side in India.

For many clinical studies involving new implants, the recruited patients often do not adhere to the follow-up required, and therefore the clinician researchers are unable to monitor the progress of the clinical trial outcome.

Other factors which are affecting the ability to do good quality research are the poor quality of writing of case files and hospital records. Good quality medical records are required for research, particularly for retrospective research. Also, many medical colleges do not employ medical statisticians. Other bottlenecks include lack of animal facilities in many newer medical colleges and lack of funding for research.

The National Medical Commission Bill was introduced by the Minister of Health and Family Welfare on 22 July 2019. The bill seeks to repeal the Indian Medical Council Act, 1956, and provide for ‘a medical education system which ensures availability of adequate and high-quality medical professionals, adoption of the latest medical research by medical professionals, periodic assessment of medical institutions, and an effective grievance redressal mechanism'. With political will to provide health for all, it is hoped that the current status of medical education will be revamped, so that the clinicians are better prepared for the twenty-first century.

5.6 Harnessing the Benefits of Biomaterials for the Country

The federal government funded multi-institutional research programmes, involving clinicians and entrepreneurs, are to be pursued with an endpoint objective of manufacturing cost-effective patient-specific implants and devices for targeted biomedical applications. Such programmes should involve careful assessment of technology and manufacturing readiness levels (TRL and MRL). New national laboratories and research institutes should be dedicated to address the clinically relevant problems in human diseases. To develop biomedical device prototypes, multi-institutional research programmes must involve clinical immersion and translational feasibility. Examples of national institutes involved in multi-institutional research programmes are Sree Chitra Tirunal Institute for Medical Sciences and Technology (SCTIMST), Thiruvananthapuram, and the Translational Health Science and Technology Institute (THSTI), Faridabad.

In the university campus, the research ecosystem must include medical hospitals for meaningful collaboration with active clinicians.

Such an ecosystem would facilitate human resource development via more interdisciplinary academic programmes with a strong curriculum, involving engineering design, manufacturing, biomaterials, biology, and data sciences.

Worldwide, the idea of patient-specific implants for musculoskeletal reconstruction is rapidly gaining acceptance among surgeons. The patient-specificity should also carefully consider the bone condition of the subject, in case of implants for musculoskeletal applications. This also changes the way in which surgeons plan and execute their surgeries. However, this paradigm has not been introduced in clinics in India to a large extent, yet.

The need for indigenous cost-effective novel dental implants with better biomechanical stability to secure the increasing oral health rehabilitation requirements, is expected to increase in the next 10 years, as the country is shifting towards having a higher proportion of elderly population. Abutment loosening is of great concern currently in dentistry, which often necessitates additional post-operative surgeries (see Fig. 5.10). This demands thoughtful design of abutment–screw interface geometry along with a rugged interlocking design to connect the components.

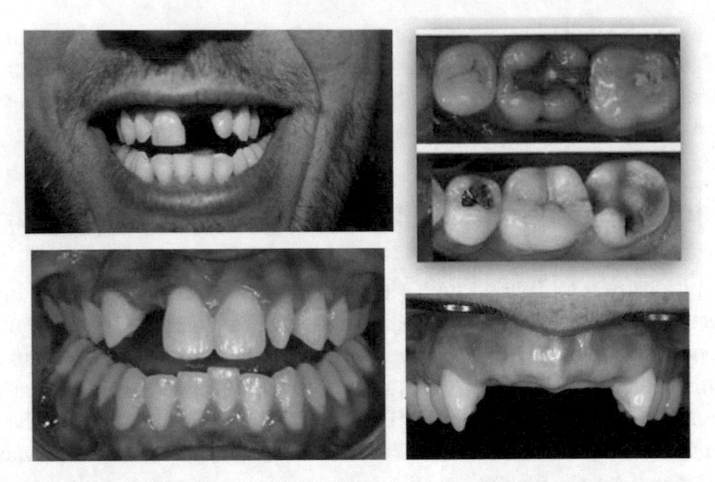

Fig. 5.10 Clinical challenges in human patients due to edentulism which requires partial/full dental restoration using implants

Patient-specific implants have shown better performance over their generic counterparts worldwide in clinics due to the precise adaptation to the region of surgery, reduced procedural times, better aesthetics and most importantly greater integration with the injury site.

It is proposed to start a national consortium, to accelerate the translation of innovative research on biomaterials science and biomedical engineering towards commercialisation of clinically validated solutions.

The national mission is to be envisioned and promulgated with leading researchers from IISc/IITs/NITs/National laboratories/industry to facilitate end-to-end total integration of manufacturing lifecycle of biomedical implants.

The mission of such a consortium would be to enable acceleration of indigenous innovation with respect to high-performance biomaterials, manufacturing/testing, materials processing technologies and biomedical devices through cross-disciplinary translational mission-mode efforts among academia, national laboratories and industries, with intensive-interactive inputs from clinicians and policy institutes.

The Government of India has shown commitment to support several technological missions. There are nine technology missions of the Office of the Principal Scientific Adviser to the Government of India. These missions are to be implemented by the Prime Minister's Science, Technology and Innovation Advisory

Council (PM-STIAC) Missions, in the areas of Natural Language Translation, Electric Vehicles, Artificial Intelligence, National Biodiversity Mission, Quantum Frontier, BioScience for Human Health, Waste to Wealth, Deep Ocean Exploration and Accelerating Growth of New India's Innovations (AGNIi). AGNIi provides a platform for innovators to scale up their market ready products by creating pathways for licensing, technology transfer and market access. This can provide an opportunity for commercialisation of indigenous biomedical implants and devices.

A challenge that is expected with respect to the uptake of indigenous implants is the mindset that indigenous implants are not reliable. Even if the indigenous implants have comparable technical specifications to the imported ones, one can expect this mindset to be widespread.

> The already-perceived notion in patients, that indigenous implants are not reliable, is to be addressed by the key opinion leaders from the medical community.

5.7 Closure

To sum up, it is essential to bring together engineers, scientists (biologists, particularly), clinicians, and industry for the development of biomaterials and implants. Many of the challenges, emphasised in this chapter (see Fig. 5.11), demand the establishment of more medtech zones with vibrant translational ecosystems in different strategic locations in India, to take the laboratory-scale innovations to market. In the next chapter, I will discuss a detailed road-map of my recommendations.

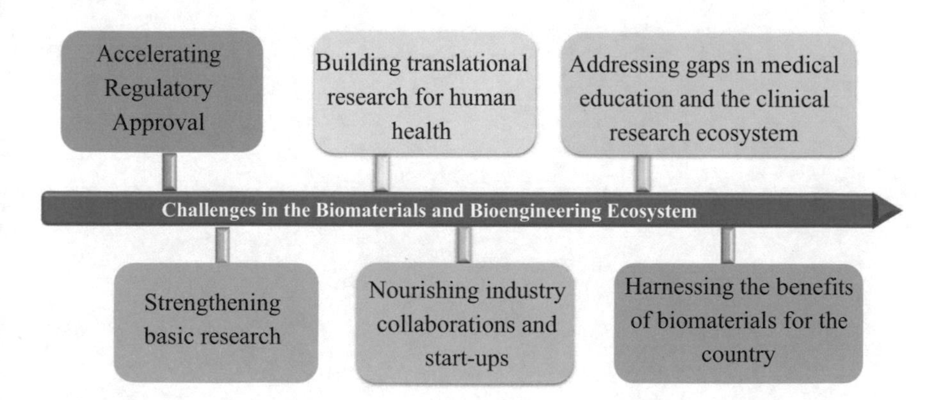

Fig. 5.11 Summary of challenges in the biomaterials and bioengineering ecosystem

Chapter 6
Recommendations

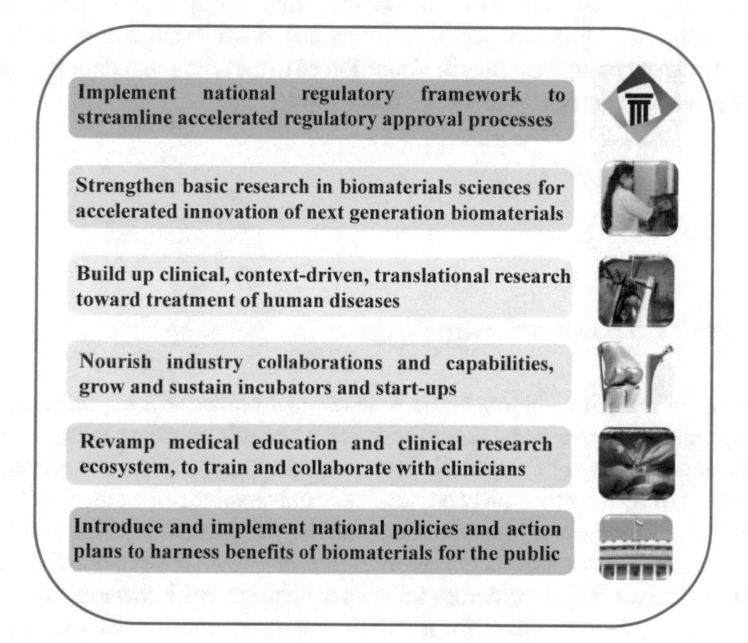

Abstract The sixfold recommendations presented here can accelerate the development of indigenous high-quality biomedical materials and implants and contribute to making the treatment of millions of patients in India affordable. On its 73rd year of independence, India is surging ahead. Right at the heart of global geopolitics, it is one of the world's best-performing human resource capitals, an emerging economy, and a software giant. Nevertheless, what about the scientific wealth of the nation? For any country that aspires to achieve faster, more sustainable and inclusive growth, science and technology need to emerge as key drivers of socio-economic development—especially so that it can cater to the healthcare needs of 1.3 billion citizens. For that, the nation needs to exploit unique innovation opportunities for its large and talented pool of scientists, engineers, clinicians, innovators, and start-up entrepreneurs, for achieving the dream of a 'healthy' India. Although the biomaterials community of India is increasingly visible for its scientific outputs and research publications, we

B. Basu, *Biomaterials Science and Implants*,
https://doi.org/10.1007/978-981-15-6918-0_6

are still struggling to execute translational research, to bridge the gap between basic and clinical research and to take laboratory scale research to the patient's bedside. In terms of translational activities in healthcare sector, significant efforts are invested for the biomedical diagnostic and extracorporeal devices, with much less visible efforts on biomaterials, implants, and tissue-engineered scaffolds.

For affordable healthcare diagnostics, the nanotechnology-based skill set, micro-fabrication capabilities, and related engineering capabilities, are particularly explored in the Indian context. Against the above perspective, it is expected that the recommendations laid down in this chapter will usher a new era in the field of biomaterials science, leading to significant translational research on implants. Many of the recommendations will be equally applicable in many of the developing nations around the world. It is emphasised that the translational research on implants demands stringent adherence to strict regulatory protocol. Summaries of relevant conferences, workshops and road-mapping meetings that contributed to the recommendations presented here are given in Appendices D and E, respectively.

6.1 The Translational Gap

The sustained growth of biomaterials sciences and biomedical engineering would certainly require intense translational research, to enable commercialisation of new implants and devices for the benefit of human healthcare. Thus, it is important to inculcate a culture of innovation in our academic universities and medical institutions, with the active support of all science and technology departments of the Government of India, state governments, and the corporate sector—to fill the existing translational gap. Various research groups across the country have to work towards coordinated, innovation-led efforts essential for scientific innovations to impact society, within reasonable timelines.

6.1.1 Science at the Local

Dramatic advances have taken place in India in the fields of biomaterials, tissue engineering, drug delivery systems, cellular engineering, and biomedical devices in recent years. Technological innovations in diverse spheres of science are driving growth: from biodegradable implants, internal fixation devices, patient-specific musculoskeletal implants to real-time sensor embedded implants. It is imperative that indigenous design, development, and manufacturing capabilities should synchronise well with translation of research knowledge into clinical practice.

6.1.2 Regulatory Impasse

A key bottleneck in the medical devices industry is the country's skewed regulatory paradigm. Let us ask why Indian-manufactured medical devices still suffer from the perception worldwide that they are low-cost and poor quality?

> The absence of well-formulated and effective regulatory mechanisms is responsible for not bringing Indian endeavours its due for performance or innovation.

While an indigenous industry has been present since the 1970s, the first attempt at regulation came about in 2006—that too, by an inappropriate notification that clubbed medical devices with drugs. The mismatch resulted in medical devices coming under the drugs regulatory body, CDSCO, with glaring loopholes that ignored the fact that devices are different from drugs and that pharmaceutical professionals cannot effectively regulate the medical devices industry.

In the last one decade, innumerable representations from industry to correct this wrong have been ignored. It is, therefore, not surprising that the large majority of Indian clinicians and health professionals continue to place their faith in 'imported' devices. As a result, imports continue to dominate, accounting for 75–95% of the Indian medical device market.

> A new and effective regulatory framework for biomaterials and implants is the need of the hour.

In order to move in the right direction, the government is planning to set up a Medical Devices Authority (MDA) for the entire spectrum in the medical devices sector, including implants. The MDA will be separate from the Central Drugs Control Standard Organization (CDSCO), and will not have any role in pricing, which will be overseen by the National Pharmaceutical Pricing Authority (NPPA). The body would comprise representatives from industry, policy makers, and active medical practitioners.

6.1.3 Steering Science and Technology

The recommendations made in this chapter are expected to align with the Government of India's new strategic initiative, the National Science, Technology & Innovation Foundation (NSTIF). This body was established to steer the management of science and technology in the country, including science education, scientific innovation, coordination, monitoring, and guiding implementations in various sectors. In interdisciplinary research programmes, young researchers from a host of disconnected disciplines—engineering to pharmaceutical sciences—have an opportunity, along with the responsibility to take on the mantle of leadership.

6.2 Priority Recommendations for Fostering Innovation and Growth of Biomedical Materials and Implants

For ambitious nation-building, six key recommendations are long overdue. Policy is at the heart of the journey towards change. I propose a road map to pinpoint directions necessary to navigate the terrain ahead.

6.2.1 Implement National Regulatory Framework to Streamline Accelerated Regulatory Approval Processes

The following recommendations are suggested to be strongly considered by the Medical Devices Technical Advisory Group (MDTAG), constituted by Ministry of Health & Family Welfare, Government of India, to advise Central Drugs Standard Control Organization (CDSCO) in July, 2019. It is felt that MDTAG should have more representatives from academia (active bioengineering researchers) and industry. Key recommendations are outlined in Fig. 6.1.

First Three Years (3)

1. Develop and enforce a medical device-specific national regulatory framework, to take into account globally validated standards provided by the International Medical Device Regulators Forum (https://www.imdrf.org/index.asp). Such a framework must recognise that implants and medical devices typically feature a wide variety of technologies, involving the material sciences, engineering, electronics, and other disciplines.

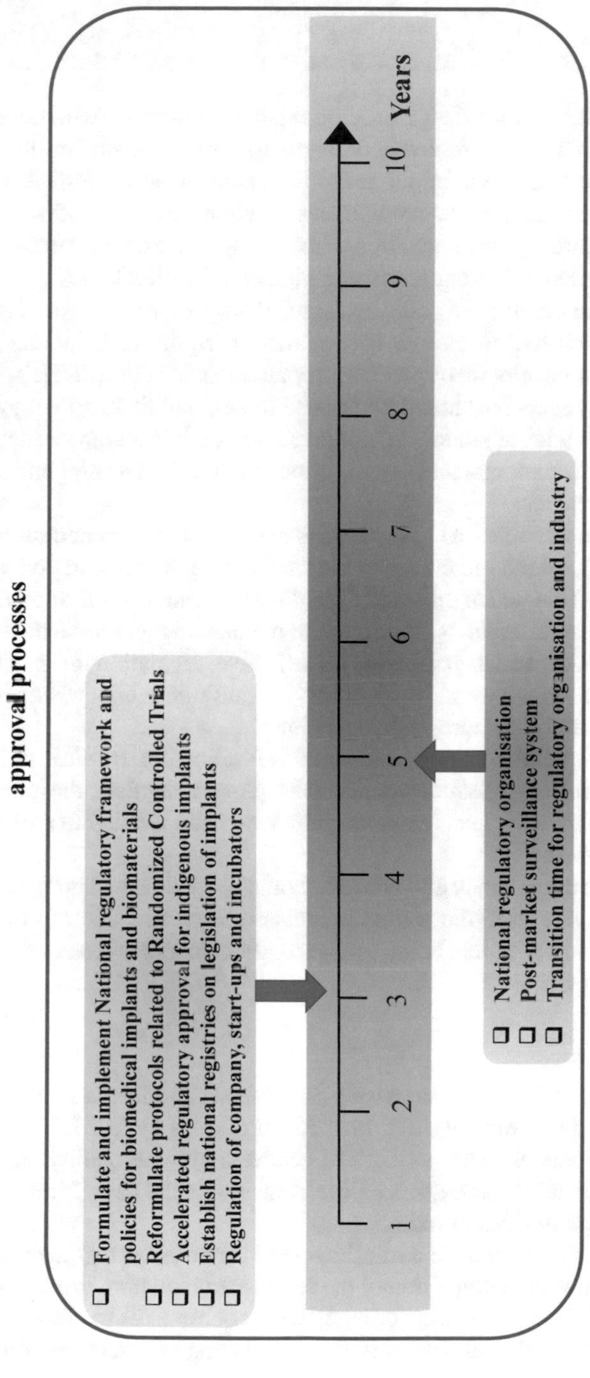

Fig. 6.1 Summary of important recommendations to develop a national regulatory framework

It is inappropriate to club medical devices and drugs into the same regulatory mechanisms and processes.

2. Develop a strong knowledge base on regulatory matters, with accessible regulatory officials who can advise on regulatory requirements during the product development stage. It is important to train students at the PhD level on regulatory requirements. Regulatory agencies should have a sense of duty to engage in dialogue regarding how regulations should emphasise their interests in protecting patients, while facilitating promising biomedical technologies.
3. Formulate an accelerated pathway for regulatory approval involving experts that can be approached to discuss the regulatory requirements at any point in the research and development process, in particular for clinical trials. It is recommended that agencies should be formed to help facilitate and organise clinical trials, in line with regulatory requirements. Also, it is recommended to simplify procedures to seek patient consent prior to clinical trials and ethical approvals for animal studies.
4. Reformulate the rules and regulations associated with randomised controlled trials (RCT), which are currently very restrictive, in consultation with medical institutions. This will significantly facilitate the clinical translational research on biomaterials and implants, in additon to maintaining an online database.
5. Build facilities and capabilities for effective regulation of medical devices through the creation of a regulatory organisation comprising of qualified professionals in the appropriate disciplines.
6. Establish national registries enabling legislation on medical devices, under national regulatory framework. Policies should regulate the use of medical devices to improve healthcare outcomes and the availability of high-quality medical devices.
7. Registration of company and product should be made mandatory, based on regulatory clearance from the national regulatory agency, similar to that applicable to international products being marketed after clearance from USA-based FDA or European CE.

First Five Years (5)

1. It is imperative that India has a well-established national regulatory organisation for biomaterials, implants, and biomedical devices by 2025. This organisation should act independently as a nodal centre to enforce quality and appropriate regulation on all the indigenous development and the use of imported implants or biomedical devices in India.
2. Account for and announce a transition time for both the regulatory organisation and industry. After completion of the announced transition time, ensure rigorous and absolute implementation of regulatory requirements to uproot all devices of inappropriate quality and organisations producing such devices from the market place.

3. Develop and enforce post-market surveillance systems (PMSS), to monitor product safety after the products are implanted or deployed.

6.2.2 Strengthen Basic Research in Biomaterials Sciences for Accelerated Innovation of Next-Generation Biomaterials

It is widely perceived that better understanding of the 'science' aspect of biomaterials development would lead to better implants and scaffolds for biomedical applications. Key recommendations are outlined in Fig. 6.2.

First Three Years (3)

1. Computational data science-based approaches, like machine learning algorithms, should be employed to accelerate the development of 'ideal' bone-mimicking material, which remained a dream for generations of biomaterials scientists. In such an approach, fine scale structural data of various (multi)-ion-doped CaP bioceramics as test data and the structural analysis of natural bone as training data can be used.

> It is recommended to effectively adapt artificial intelligence (AI) approaches and/or machine learning (ML) algorithms for better predictability of clinical performance of biomaterials/implants.

2. Taking forward significant interest in 3D and 4D Bioprinting in the next few years, an exciting recommendation would be to formulate and implement '**Space Bioengineering**' as a flagship national programme, in line with ISRO's human spaceflight programme. Such a programme would enable biofabrication of programmable self-assembly of 3D tissue and organs constructs in micro-gravity culture conditions. This requires rational design and manufacturing of 3D bioprinters with significant on-ground testing and evaluation of spaceworthiness. The scientist can plan to conduct on-ground experiments to use 3D bioprinting to develop scaffolds with spatio-temporal control of molecular and physical signalling for the newly regenerated tissues to mature, to whole organs for transplantation.

> Once successful, this will place India as the only fifth nation to launch such an ambitious space bioengineering program.

3. Synthetic biopolymers and natural soft biomaterials are by far the most widely investigated in India and globally. However, structure–property correlation of soft

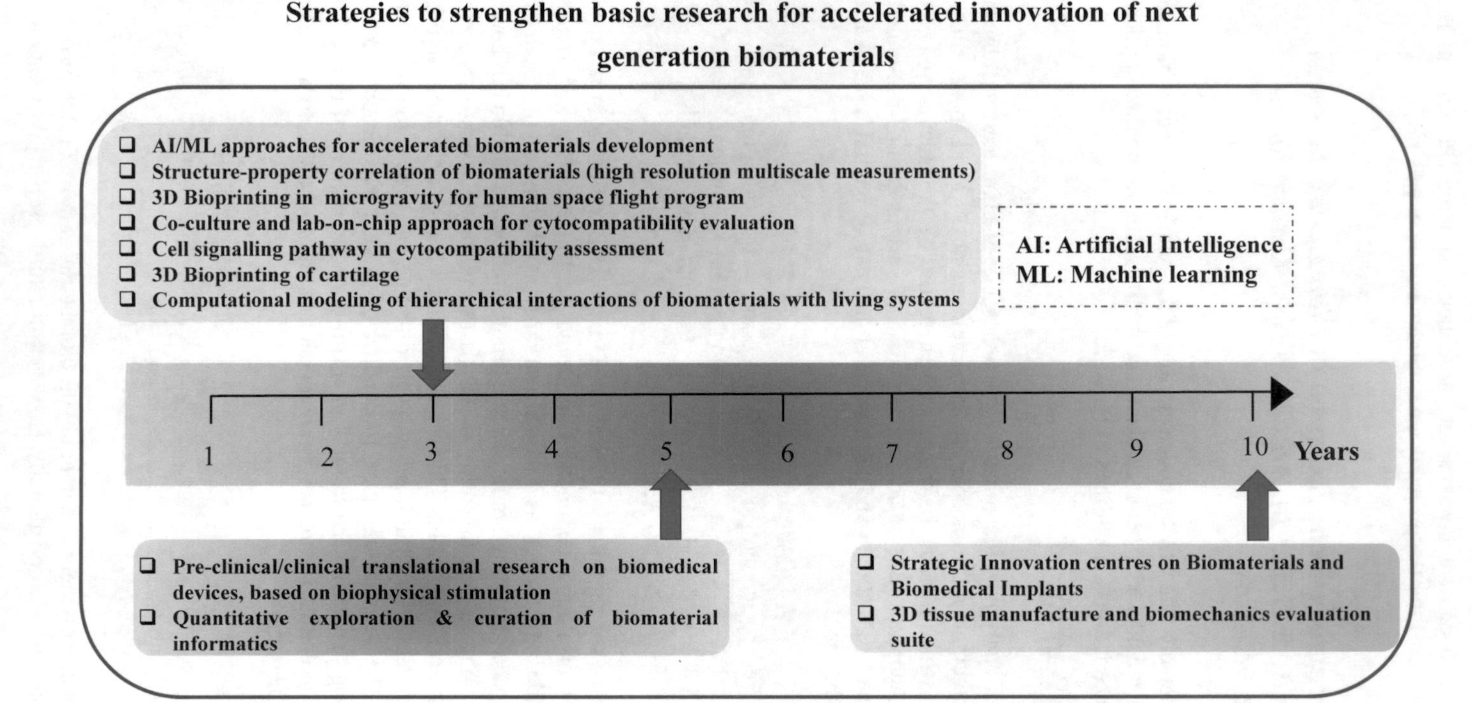

Fig. 6.2 Summary of the major recommendations to strengthen basic research in biomaterials science

materials remain largely unexplored. It is recommended to explore the fundamental understanding of macromolecular structure/viscoelastic property correlation of polymeric biomaterials using a diffusive wave spectroscopy (DWS)-based micro-rheology approach, together with small angle X-ray scattering (SAXS). Such combinatorial analysis allows one to probe into the origin of the viscoelastic property variation in terms of structural changes due to the complex interactions among macromolecules.

4. The structure–property correlation of many of the implantable bioceramics is not well established. It is recommended to use a synchrotron facility, i.e., extended X-ray absorption spectroscopy (EXAFS), to record high resolution data for compositionally tailored CaP-based bioceramics, one of the widely investigated inorganic biomaterials. This, together with computational techniques, like molecular dynamics simulation, density functional theory (DFT), can be used to precisely determine the dopant distribution in energetically stable crystal structures of doped CaP biomaterials.
5. Regarding cytocompatibility evaluation, the first recommendation would be to co-culture endothelial cells with target tissue-specific cells, as such an approach facilitates better maturation of engineered tissues by enhanced capillary formation in extended timescale. The second recommendation would be to adopt the lab-on-chip approach, wherein *in vitro* experiments are carried out in designed biomicrofluidic devices under physiologically relevant shear.
6. It is also recommended that researchers develop an in-depth understanding of the cell signalling pathways involved in cell functionality changes for implantable biomaterials. For advanced biomaterials-based nanotechnologies, cellular and tissue therapeutics should be focused on the fine tuning of cellular behaviour. Also, there should be development in advanced delivery systems for drugs, proteins, peptide-based products, and gene therapeutics. This requires extensive collaboration with biologists.
7. With reference to 3D printing, it is recommended to formulate a universal binder to print a wide spectrum of implantable biomaterials, as well as to develop insightful understanding of the binder-material interactions, preferably using synchrotron studies.

> Currently, most of the commercial 3D printers can be used to print standard materials using the company's patented binder.

This severely restricts research on 3D printing of patient-specific implants using new biomaterials and material-specific binder.

8. With respect to **cartilage tissue engineering**, a biomaterial, acting as a cell carrier for cartilage repair, should serve as a temporary biodegradable scaffold that permits synthesis of extracellular matrix from delivered cells, while leading to a close fit attachment to the perimeter of the cartilage defect. Super macroporous (0.1–100 μm) polymeric gels can offer great potential as scaffolds for

cartilage tissue. An *in situ* gelling system can help to overcome the problem of having a good fit of the scaffold to the perimeter of the cartilage defect (less invasive delivery form). Chondrocytes need to be maintained in the rounded hyaline form (with suitable growth factors).

9. With respect to optimisation of a material and therapy for bone tissue engineering, laser/electric field stimulation can guide mesenchymal cells towards osteogenesis, contributing to bone growth. Synergistic interaction among the electric/magnetic field stimulation and biomaterial substrate properties can guide stem cell differentiation through different lineages. The outcome of such research programmes will have significant relevance for the emerging field of **Bioelectronic Medicine**. Another approach could be to examine the efficiency of highly intense laser sources in order to establish a dynamic poling of electroconductive composites, and subsequently study osteogenesis on such laser-treated surfaces.

First Five Years (5)

1. Similar to cell culture experiments, the antibacterial properties of biomaterials are mostly evaluated using a single pathogenic strain type (*S. aureus, S. epidermidis, E. coli*, etc.). The competition involved in interactions among prokaryotes and eukaryotes are much less explored in the biomaterials community and cannot be assessed by studying their respective interactions with biomaterials, in isolation. The recommendation would be to explore this aspect, prior to pre-clinical study.
2. In the context of electric field stimulation of cells on biomaterials, it is recommended to adopt quantitative modelling approaches, using molecular dynamics (MD) and first principle calculations. Deeper knowledge in this area would help to develop biophysical stimulation devices for regenerative engineering applications.
3. The bench-to-bed translation of the concept of 'electrical field stimulation of stem cell niche' in the context of tissue engineering and regenerative medicine is another aspect that needs further investigation. Although a multitude of *in vitro* studies are being reported in literature, translational research would require more pre-clinical studies, with microelectronic miniaturised modules to deliver electrical stimulation, followed by thoughtful clinical trials.
4. It is recommended to establish a mathematically rigorous cross-disciplinary framework that will allow a systematic quantitative exploration and curation of critical biomaterials informatics to objectively drive the innovation efforts within a suitable uncertainty quantification framework. Additionally, one needs to consider multiple physics-based approaches and potential coupling between them in extracting this core knowledge (e.g., multifunctional properties needed for biomedical applications). This will enable a rigorous foundational framework that allows objective extraction and fusion of the incomplete and uncertain materials knowledge, to address the central impediment in streamlining and enhancing the efficacy of current biomaterials innovation efforts.

First Ten Years (10)

1. What is needed in the next one decade is to develop strategic innovation centres to support the faster development of biomedical devices, which are to be guided by an extensive network of research groups across bioengineering/materials science/mechanical engineering/biological sciences/chemical sciences/computational data sciences in academic universities, medical institutions, and national laboratories. These centres should facilitate long-term strategic research programmes involving basic scientists, doctoral students, and post-doctoral staff, regulatory consultants, for designing and prototyping of biomaterials-based medical devices. Another recommendation would to establish biomedical device testing facilities in those strategic innovation centres, as mentioned above. Currently, good materials testing and characterisation facilities are available in major academic institutions.

> High-throughput manufacturing and accelerated biocompatibility testing should be the focus of these futuristic strategic innovation centres.

2. More multidisciplinary research efforts in the area of 3D (Bio) printing are to be invested, involving analysis of complex anatomical features, clinical imaging (CT or MRI scans) algorithm and printer development. Specific attention is to be paid to optimisation of bioink rheology printing parameters, gelation kinetics, structural and mechanical stability, and biodegradation of soft polymeric hydrogels. Printing of a variety of materials should be possible with high spatial control and resolution, through multiple printing heads (co-axial extrusion) using different techniques such as heated, non-heated, pressure-assisted, and extrusion combined with a laser unit for direct polymerisation. Further applications to be pursued are drug development and toxicity analysis, scaffold-based tissue engineering, automated tissue-based *in vitro* assays for clinical diagnostics, and modular tissue assembly.
3. It is recommended to build a 3D tissue manufacture and biomechanics evaluation suite. The suite should allow high-throughput culture of a variety of tissue and acellular products, such as heart valve tissue, blood vessels, ligament/tendon bone, and cartilage. It should also allow testing of acellular biomaterials, such as hydrogels, polymers, elastomers (rubbers), ceramics, metals, composites, etc.

6.2.3 Build Up Clinical, Context-Driven, Translational Research Toward Treatment of Human Diseases

The major recommendations to build a stronger and more effective translational research ecosystem are summarised in Fig. 6.3.

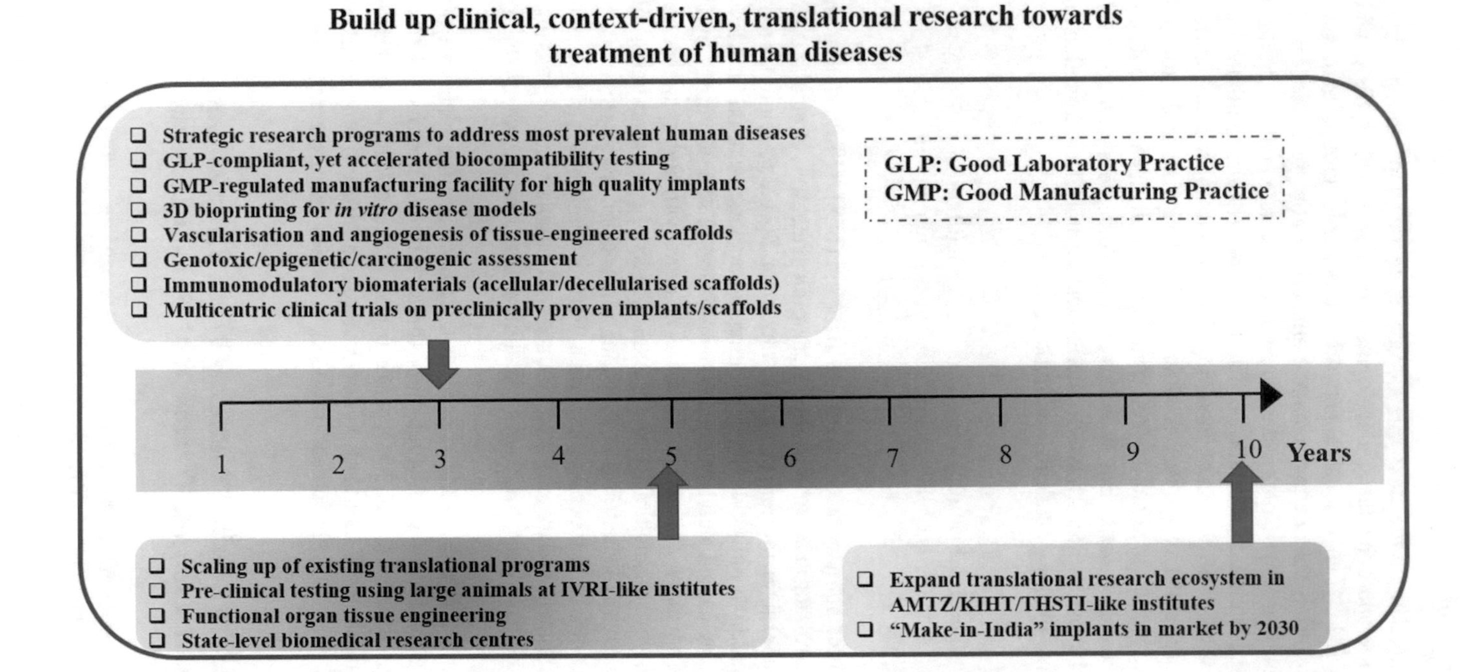

Fig. 6.3 Recommendations for enhancing translational research ecosystem on biomaterials and implants

First Three Years (3)

1. It is recommended to launch strategic research programmes on the most prevalent human diseases. The national societies like Society for Biomaterials and Artificial Organs (SBAOI) or Biomedical Engineering Society of India (BMESI) or Society for Tissue Engineering and Regenerative Medicine India (STERMI) together with National Academy of Medical Sciences (NAMS) are suggested to conduct critical analysis of clinical cases registered across hospitals/clinics to arrive at the ranking of the most prevalent human diseases. The funding agencies can call for proposals on a priority basis to address the unmet clinical challenges.
2. In the context of biocompatibility testing, the researchers mostly report the cytocompatibility results, analysed using a host of complementary molecular biology assays. It is equally important to assess the genotoxicity of the nanobiomaterials in a size- and dose-dependent manner, using comet assay, single-cell gel electrophoresis or micronucleus (flow cytometry) techniques. The quantitative analysis, together with the mechanism of DNA fragmentation, has large significance for epigenetic changes. This obviates the necessity of evaluating genotoxic, epigenetic, and carcinogenic effects, to improve the clinical acceptability of the material.
3. It is recommended that the researchers invest future efforts to explore the host response modulation of immunomodulatory scaffolds (acellular or decellularised), using the concept of regenerative immunology (phenotypic plasticity of immune cells, or by targeted co-delivery of immune cells with the organ-specific functional cells).

> Immuno-modulation should emerge as a therapeutic discipline which can cater to the needs of transplantation, transfusion medicine, tissue engineering, regenerative medicine, and implantation.

4. Another important recommendation is that vascularisation phenomena require significant attention of the researchers pursuing tissue engineering and regenerative medicine.
5. In the context of 3D bioprinting, it is recommended to focus first on development of *in vitro* disease models, which could be used by Indian pharmaceutical companies for *in vitro* drug screening.
6. Establish more **GLP-compliant research facilities**, such as those at CSIR-IITR and SCTIMST, at strategic innovation centres. It is recommended to follow the International Organization of Standardization (ISO) protocols for evaluating biocompatibility of biomaterials or devices using animal and human cell cultures.
7. Another equally important recommendation is to establish **GMP-compliant manufacturing facilities**, in strategic innovation centres for translating research outcomes to marketable implants/devices. Specific standards for safety, Quality Assurance (QA)/ Quality Control (QC), and ethical norms need to be enforced in all these facilities.

8. Knowledge transfer and training of a well-trained human workforce with multidisciplinary capability for creating an innovation ecosystem. The workforce for tomorrow must be exposed to the diverse multidisciplinary knowledge of medicine, biology, chemistry, materials science, engineering/manufacturing, etc. This would allow them to collaborate across the silos, where researchers often confine themselves to traditional knowledge domains.

First Five Years (5)

1. The scaling of the existing translational programmes with more effective interfacing among clinicians, engineers, and biologists is recommended. The emphasis should be on developing cost-effective healthcare solutions of high quality. One example of a recently opened centre is the 'Centre for Engineering in Medicine' at IIT Kanpur, which will offer Masters and PhD level courses.
2. Establish large animal facility to cater to significant pre-clinical biocompatibility study of the biomaterials. Currently, Indian Veterinary Research Institute (IVRI), with which is headquarter at Izatnagar, UP, has five campuses at different locations in the country, namely Pune, Kolkata, Palampur, Mukteswar, and Bengaluru. Many of these institutes have large animal facilities primarily to cater to investigate prevalent animal diseases for better healthcare of the animals as well as for drug/vaccine development.

> India should have more specialised institutes for pre-clinical testing of implants in large animals under the broad umbrella of IVRI.

3. In the field of whole functional organ tissue engineering, it is recommended to pursue research on cartilage, bone, bladder, trachea, heart valve, heart, lung, nerve, skin and kidney, besides whole tissue reconstruction. Of note, technologies for connective and skeletal tissue like critical bone defects, wound care management, and vascular grafts are already maturing in India.
4. It is advised to develop biomaterial and biomedical research programmes, which can be funded by the ministry of health or the ministry of science and technology of regional states in India. After these state-level centres demonstrate satisfactory progress, those units may be enabled to evolve into full-fledged national research institutes.

First Ten Years (10)

1. A long-term recommendation would be to establish a vibrant translational research ecosystem, comprising of a group of institutions of eminence as members of a consortium, which transfer technologies at TRL 4–5 level to an industrial partner (start-ups, MSME, or larger industrial partner). This should essentially be followed to establish SCTIMST/THSTI–like translational research institutes in the area of biomaterials and implants. It is important to establish

newer translational research institutes that should have R & D facilities that can be utilised for large animal studies on biomedical implants and devices.

The new translational institutes should facilitate multi-centric clinical trials on pre-clinically proven implants and biomaterials.

2. The majority of indigenous biomedical devices/implants should have a '*Make-in-India*' label by 2030.

6.2.4 Nourish Industry Collaborations and Capabilities, Grow and Sustain Incubators and Start-Ups

One of the important lacunae in the entire ecosystem is the lack of strong presence of indigenous industry to complete the cycle of 'science-to-product' innovation. To this end, the major recommendations are outlined in Fig. 6.4, and more details are given below.

First Three Years (3)

1. Establish strategic policy centres to conduct market research on biomaterials and implants. The dynamic changes in implant market should be carefully studied, and the market research results should be accessible to industries, MSMEs, and start-ups.
2. Strengthen and expand KIHT/AMTZ–like ecosystems across different strategic locations in India to create strategic innovation centres to facilitate transfer of clinically validated technologies from academia/national laboratories to established industry. Such centres should develop the capability for commercial-scale production of biomedical-grade materials such as hydroxyapatite powders/blocks/granules, Ti6Al4V/stainless steel/CoCrMo blanks, UHMWPE blocks, and biodegradable polymers (PLA, PLGA, etc.).

The unavailability of biomedical grade powders or solid blanks for proof-of-concept studies often restricts research in academia and national laboratories.

3. KIHT/AMTZ group of institutions should maintain a national registry of medtech innovations from across the country. This should be made available to companies, which can then analyse the commercialisation potential of the innovations and take the promising ones to market.
4. The healthcare sector should have significant investment from the government for manufacturing of biomedical materials/implants, and for an appropriate regulatory framework for the sector; it thereby should minimise the import of substandard or even recalled devices. Competitive cost-structure analysis of newly

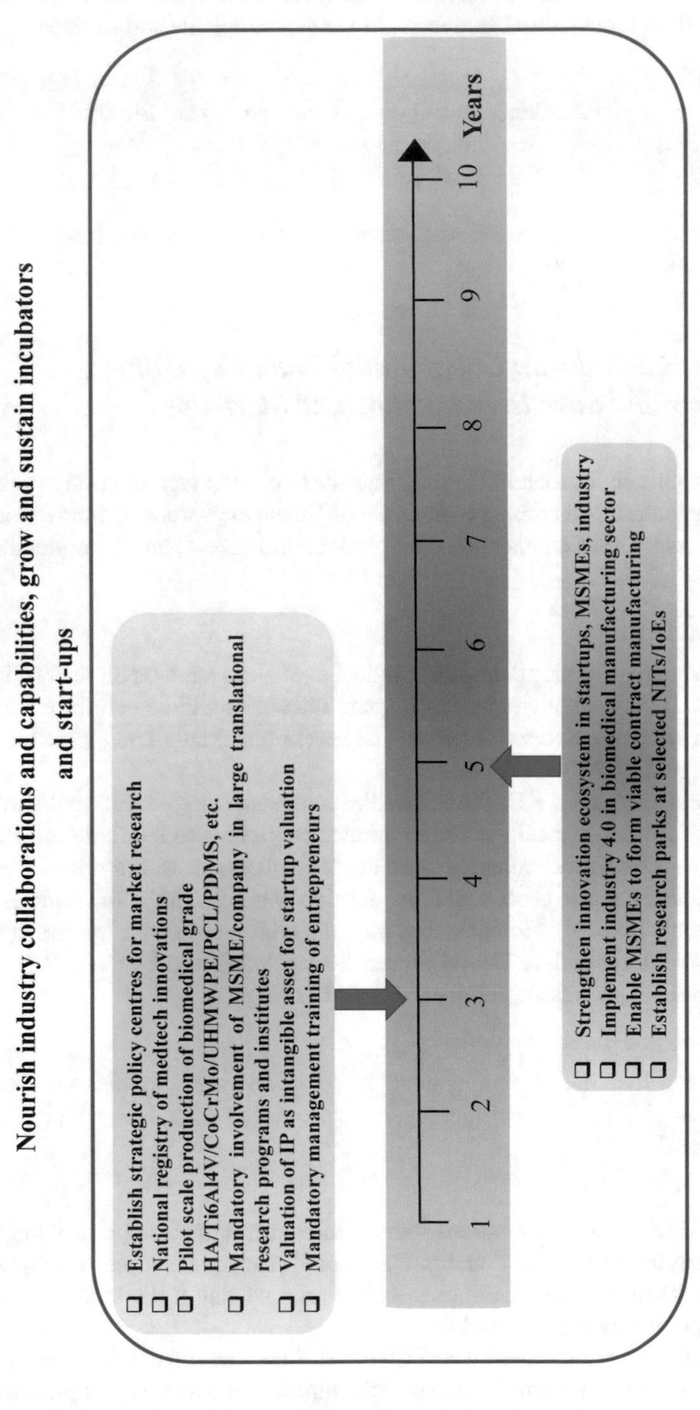

Fig. 6.4 Key recommendations to nurture and nourish industry collaborations

indigenous implants and materials with respect to similar products currently sold worldwide is warranted.

5. Promote industry-focused clinical excellence as a driver of translational research and involve industry, right from TRL-1 stage, while conceiving translational research programmes at strategic innovation centres. This demands commensurate funding mechanisms that strongly emphasise collaborative research with at least one mandatory industry partner with tangible incentives, to ensure impact beyond publications.

The institution–industry teams should carry out due diligence to establish the technologies' assets and liabilities and evaluate their commercial potential.

6. Mandatory management training for biomedical entrepreneurs, so that they learn about strategy, branding, product positioning, assessing risk factors, development of business and commercialisation plans and marketing, sales and distribution strategies of biomedical implants and devices.
7. Industry experts to participate as adjunct faculty in professional education (e.g., PhD programme in Biomedical Engineering) to help build a knowledge base on market opportunities and needs and also commercialisation of products and technologies.
8. Create incentives to attract indigenous entrepreneurs, recognise incubators as for-profit companies, and enhance the number of start-ups (both private and public institutional). Engage with the next generation of researchers to develop an entrepreneurial spirit and awareness, which is very important for developing the next generation of start-up founders.
9. An IP training certification of all innovators before receiving funding exposes them to IP-related litigation procedures and also protocols with freedom to operate analysis. Valuation of IP is an intangible asset for start-up valuation.

First Five Years (5)

1. Establish and strengthen the innovation ecosystem in start-ups/MSMEs/industry. Promote more standard contract agreements between industry and academia to accelerate engagement with single or multiple universities. It is recommended to introduce an optional tenure of biomedical researchers in industry for 6 months every 2–3 years in their career. Researchers from academia/national laboratories should be allowed to act as part-time consultants for industry R&D programmes. Without innovation, they may not be able to generate patents, and as a result they might fail to compete with international big players. Incubators must have specific skills to assess the commercial potential of technologies, and incubatees should be mentored by a pool of mentors for innovation and translation.
2. Enable Indian MSMEs to form viable contract manufacturers with market penetrating strategies by selling products to established companies, such as Meril

Healthcare, Stryker, Smith and Nephew, and Zimmer. For example, the top global orthopaedic manufacturer companies are consuming 1.5–2 million femoral heads a year. Also, the dental market in India presents a good opportunity, since there are not even a handful of local manufacturers yet.
3. Improve the country's knowledge of the manufacturing process and its effects, through courses on automation, information from sensors, data exchange (Fourth Industrial Revolution of digitisation, i.e., **Industry 4.0**) and on minimising environmental impact.
4. Establish research parks at selected NITs and institutes of eminence to foster incubation of start-ups; build large research hubs with state-of-the-art facilities in the area of advanced manufacturing (e.g., biomedical prototyping, etc.). Recently, the Indian government, under the Startup India initiative for Higher Education Institutions (SIHEI), sanctioned the establishment of the research parks at IIT Delhi, IIT Kanpur, IIT Hyderabad, IIT Kharagpur, IIT Bombay, and IISc Bangalore, with a total sanctioned cost of 575 crores.

6.2.5 Revamp Medical Education and the Clinical Research Ecosystem, to Train and Collaborate with Clinicians

Another equally important bottleneck in the clinical translational research ecosystem is getting significant involvement of clinicians. Some major recommendations to revamp education, training, and research in medical institutions are summarised in Fig. 6.5. Many of the important points are discussed below.

First Three Years (3)

1. Develop and implement a robust innovation ecosystem in medical and dental institutions, so that clinical research should be seen as the key way of ensuring that the unmet healthcare needs are addressed. Medical institutions should have an 'Intellectual Property and Technology Licensing Cell' to facilitate protection of medtech IP, and to enable technology licensing.

> It is recommended that final year MBBS/BDS or MD/MS/MDS students are involved in research projects, involving clinical studies or trials of biomaterials/implants/scaffolds in human subjects or pre-clinical studies in large animal models.

2. Start funded masters programmes at medical colleges or non-clinical institutions, where students can be awarded an M. Sc. or M. Tech. degree, if they do good quality research for two years after their MBBS or BDS/MDS.
3. Design courses on research methodology, as part of the medical curriculum in India. Faculty from good research organisations can start online courses (NPTEL programme) that can be made available in India for guiding others. They can teach

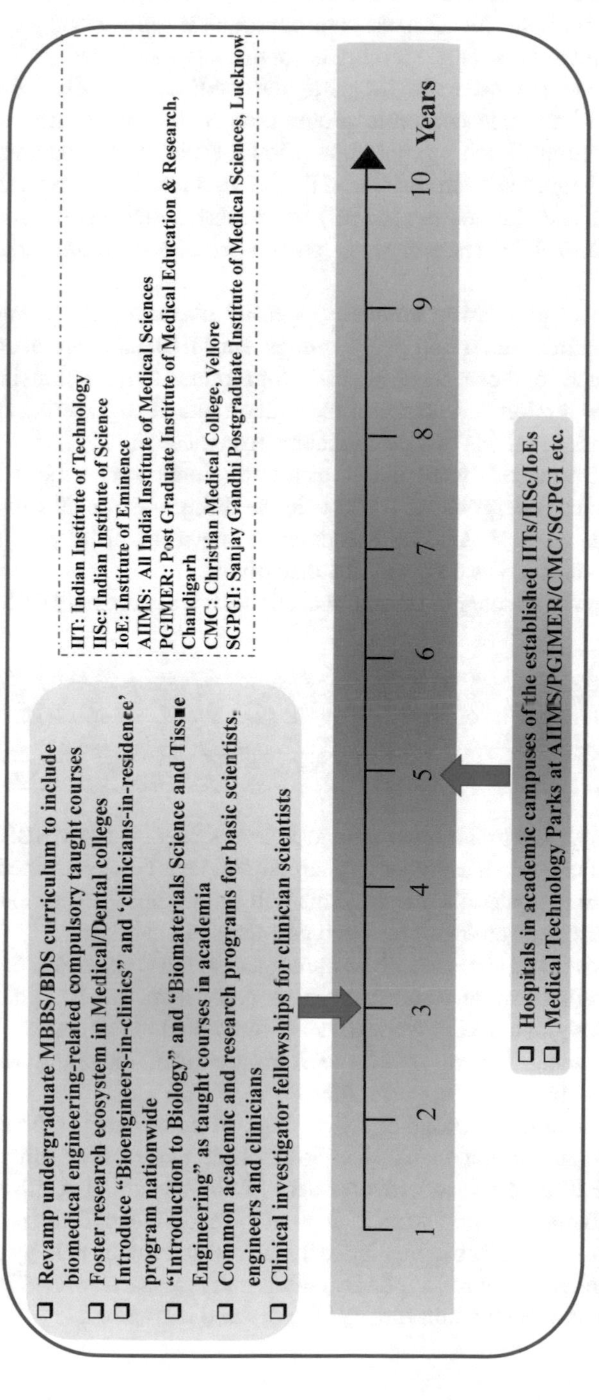

Fig. 6.5 Important recommendations to revamp medical education and innovation ecosystem in medical institution

case studies and ensure that Indian research gets adequately presented in these courses. Periodic certification based on continual learning is suggested to ensure that the quality of students' learning is continually monitored.

4. All of the major medical institutes, particularly AIIMS, PGIMER, SGPGI, CMC, etc., should have one academic programme in the broad area of 'Biomedical Engineering'. Such a programme should facilitate the internship of engineering undergraduates from IITs/NITs and also foster the research ecosystem for medical undergraduates (MBBS) or masters (MD/MS) or advanced postgraduate (DM/MCh) researchers to pursue projects on biomaterials/implants, bioelectronics, etc.
5. Formulate and popularise nationally, specific internship programmes such as '**bioengineering immersion programmes**' for MBBS students at an early stage in their career, 'clinician-in-residence' programmes to train clinician-scientists, '**clinical immersion**' programmes for undergraduate engineering students, and 'bioengineers-in-clinics' for PhD students from academia.
6. Biomaterial science is to be taught as a core compulsory subject in academic institutions for undergraduate students in the discipline of materials science and engineering. In most IITs, an introductory course on 'Biology' is taught for undergraduate students across all the disciplines.
   Importantly, IIT Kharagpur has announced that it will start an MBBS programme from 2020.

> The establishment of hospitals within academic campuses is certainly a step forward towards a stronger translational clinical research ecosystem.

7. It is also suggested to introduce a major/minor system in the MBBS curriculum, for related fields, such as major in pharmacy/biotechnology/biomedical science and minor in biomaterials science. This will inspire next-generation researchers to address cross-disciplinary research problems in healthcare.
8. Common academic and research programmes for basic scientists, engineers, and clinicians will allow to create a research ecosystem, where clinicians, scientists, and design engineers work in cross-functional collaboration. The clinical community should be engaged through programmes, including MD-PhDs and post-doctoral fellowship opportunities.
9. Provide clinical co-investigator fellowships or consultancy schemes for practicing clinicians, which enable them to dedicate part of their time for research activities. Full-time research involvement of clinicians will positively influence the translational research impact. It is also recommended to introduce more adjunct faculty positions in medical colleges, which can be held by non-medical faculty to enhance cross-disciplinary collaborative research work. This will allow closer interactions with non-medical faculty and researchers.

<u>**First Five Years (5)**</u>

1. Among all the IITs, or institutes of eminence, IIT Kharagpur has started a medical science and technology master's programme for MBBS students. A medical hospital is currently being constructed at IIT Kharagpur, and the same has been planned at IIT Kanpur. Also, several biomedical researchers at IIT Delhi collaborate with clinicians at AIIMS. Two recommendations are (a) establish more such hospitals in the vicinity of other institutes to facilitate clinical translational research, including rural areas where the topic of research is relevant to this setting and (b) appoint clinician-scientists in dedicated clinical research laboratories in all these hospitals.
2. Establish medical technology parks in reputed medical institutes to host Indian biomedical implant companies. These parks will also enable to initiate interdisciplinary research to address clinically relevant problems (involvement of clinicians, entrepreneurs, and basic scientists is necessary).

> An 'ecosystem' needs to be developed at centres of excellence at various medical institutes like AIIMS, PGIMER, Chandigarh; CMC Vellore; SGPGI Lucknow; on advanced training for clinicians, e.g., use of robotics, biomedical instrumentation, implant testing etc.

3. The MBBS curriculum should be revamped to introduce 'hands on' training, biomedical engineering concepts, and research methodologies. Both the numbers of research-capable doctors and their skill sets are expected to improve, if these changes are implemented.

6.2.6 Introduce and Implement National Policies and Action Plans to Harness Benefits of Biomaterials for the Public

All the recommendations proposed above have to align with the national policies under the broad framework. A number of vision-related strategies are summarised in Fig. 6.6, and those are briefly discussed below.

<u>**First Three Years (3)**</u>

1. An important recommendation would be to launch a new mission, 'Bioimplants' to accelerate innovation of biomedical implants for healthcare, which can be mentored and supported by Prime Minister's Science, Technology, and Innovation-Advisory Council (PM-STIAC). The purpose of this national mission would be to implement several of the recommendations suggested in this monograph. The specific committee to be constituted under this mission would be representatives from a number of related federal ministries of the Government of India (Ministry of Science and Technology, Ministry of Skill Development and

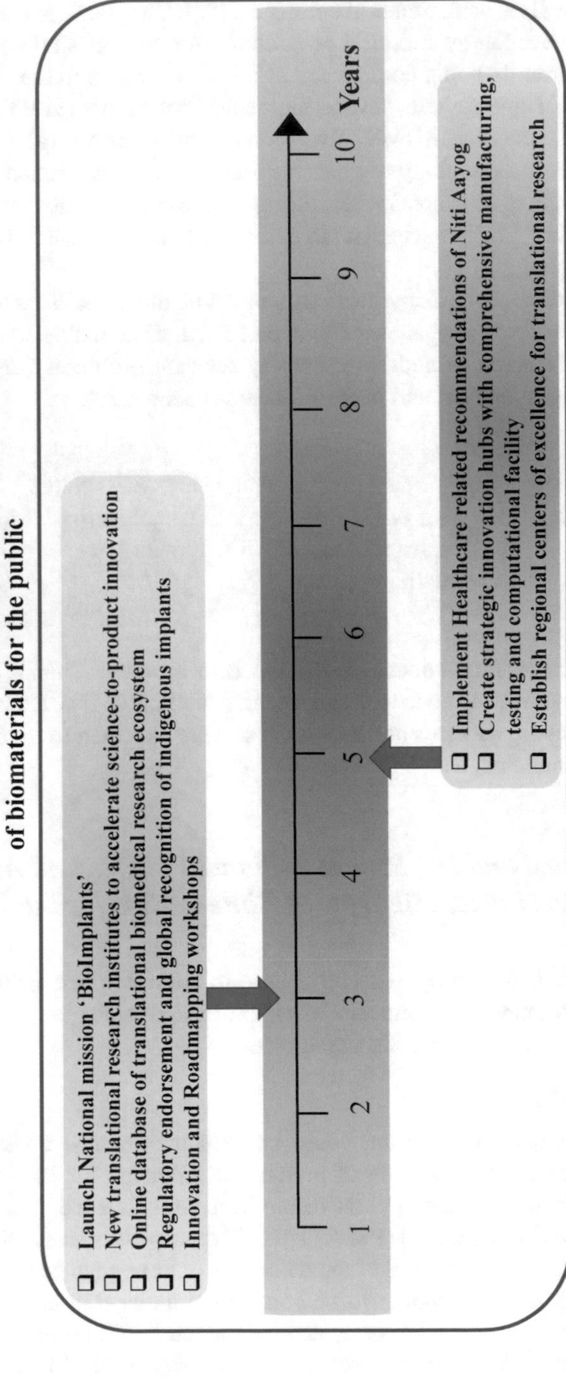

Fig. 6.6 Major recommendations to introduce new national policies

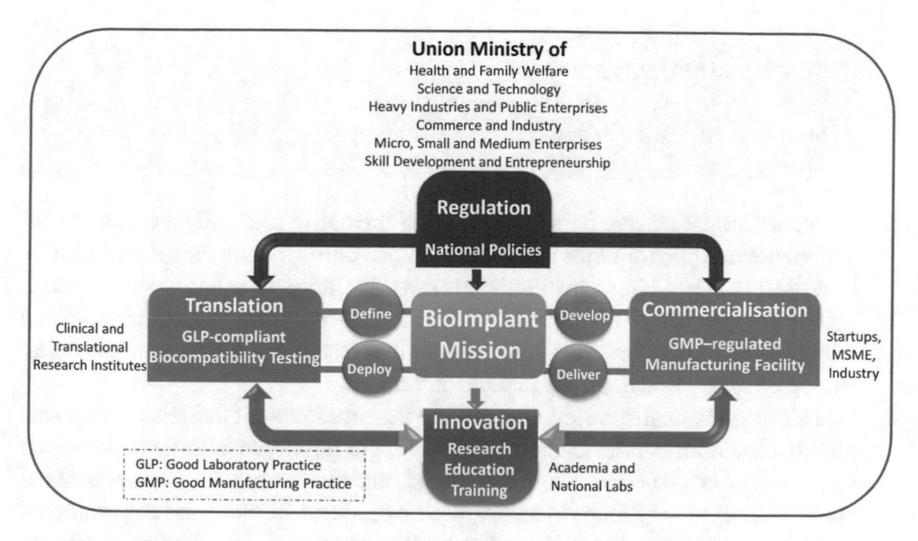

Fig. 6.7 Final recommendation to implement national policy for accelerated implant manufacturing and deployment, enabling the mission 'Healthy India'

Entrepreneurship; Ministry of Micro, Small, and Medium Enterprises; Ministry of Commerce and Industry; Ministry of Heavy Industries and Public Enterprises, and Ministry of Health and Family Welfare).

> The objective of this national mission will be to formulate effectively and regulate policies to manufacture and to deploy high quality, yet affordable biomedical implants and devices for a large cross-section of Indians (see the model proposed in Fig. 6.7).

The participation of all the ministries are necessary in view of the technological nature of innovation, regulatory compliance, and large commercialisation potential in India and other countries, as well as societal implications. The above-suggested mission can also implement subsidies for indigenous implants and biomaterials, and interface regularly with insurers for refining insurance schemes. One of the long-term objectives of this proposed mission would be to grow the global reputation of Indian biomedical implants, devices, and translational research.

2. Establish new translational research institutes and launch multiple research programmes to accelerate science-to-product innovation, and to address unmet national needs for musculoskeletal, dental, cardiovascular, neurosurgical, and urological applications. All these programmes should be conceived and constantly guided by clinicians from conception (problem definition) to validation (clinical trials) to marketable products (commercialisation).

The federal funding agencies should prioritise funding of the proposals with an objective to conduct human clinical trials, while translating research outcomes to marketable products.

3. Create a comprehensive online database on the entire biomedical ecosystem of active research programmes with details of pre-clinical and clinical trial results, as well as licensed biomedical implants/devices (indigenous/imported) having a unique device identifier code and cost. The database should be regularly updated with patient's comments, revision surgeries and recalled products, etc. The portal should be accessible to all the patients in India.
4. Constitute a national committee to oversee the promotion of indigenous implants and to disseminate relevant information to the public so that their perceived concerns can be dispelled. The committee should also work with relevant medical societies and patient groups to ensure that the products are endorsed and promoted through print/electronic media and at healthcare events. The medical fraternity can review the clinical data extensively before giving their endorsement.
5. Promote a nationwide strategy of the 'innovation stream', where research is fast-tracked by the 'fail fast, learn fast' approach—a cultural shift, which makes it acceptable that innovations can undergo failures and one should learn from failures and improve.
6. Innovation and road-mapping workshops—activity focusing on scientific exploration of advanced biomaterials, challenges, foresighting, and development of innovation strategies. One such platform can be National Frontiers of Science (NATFOS), which is supported by the office of the Principal Scientific Advisor to the Government of India, since 2018. Active researchers should put in all efforts to popularise this socially relevant research area through outreach activities like public lectures, internships, and workshops, including laboratory demonstrations in engineering colleges/high schools, to stimulate young minds

First Five Years (5)

1. Implement the broad recommendations of Niti Aayog, the national policy body, for health technology assessment as a basis for informed decision making and priority setting, in corroboration with the World Health Organization's systematic reviews of the medical devices for use in priority populations.
2. Establish regional centres of excellence in different strategic locations of India with evolving ecosystems, for better coordinated translational research, with faster and regular interaction among a group of researchers from academia, national laboratories, and industry/start-ups. An important mission of such centres should be to provide necessary infrastructure to assess safety and efficacy of medical devices, particularly on long-term clinical data, related to patient follow up in post-operative timelines.
3. Create strategic innovation hubs with comprehensive manufacturing, testing and computational facilities, which is essential for the accelerated development of

biomedical devices from concepts to marketable products. The innovative business model of the biomedical entrepreneurs should support such an ecosystem. Funding agencies and policy makers should introduce commensurate schemes.

A final recommendation to implement national policy for accelerated implant manufacturing and deployment, enabling the mission 'Healthy India' is depicted in Fig. 6.7.

Chapter 7
Closure

Abstract This chapter portrays the author's futuristic vision of a bright future of biomaterials science and implants in India. This chapter is followed by the genesis of his set of recommendations, laid down in the previous chapter, as well as a few important pedagogical aspects for the young researchers.

This monograph will, hopefully, stimulate an accelerated journey of discovery, together with manufacturing and commercialisation of biomedical devices and materials, in a mutually complementary manner. The goal is to deliver what India demands and aspires to be: an international leader in the field of biomaterials and biomedical implants and devices. Such an ambitious goal can be achieved by leading by example, planning strategically, providing the necessary resources and streamlining the innovation, regulatory, and commercialisation process. India must realise that

B. Basu, *Biomaterials Science and Implants*,
https://doi.org/10.1007/978-981-15-6918-0_7

high quality, yet affordable implants are needed to meet society's needs, and those of the booming medical tourism industry.

Health is at the very core of society's fundamental strengths. Therefore, it is recommended that clinicians, scientists, policy makers, and industries become inspired by looking at some of the successful ecosystems around the globe to develop a scientifically and socially conscious platform in India, to foster a robust innovation ecosystem within clearly defined and ethically regulated boundaries.

Despite the many scientific and medical subfields and subspecialities, the field is united by common challenges and goals. These unifying factors make a collective fraternity among professionals in the field, and should encourage collaborative efforts in the right direction. This text speaks to the moment and beyond. It is addressed to all consumers of science and medicine, critical as well as lay. It is a first of its kind road-map that can assist India to take a leading role in the flight of domestic frontier science and medicine, and regulatory policy, from the realm of the institutions to the markets catering to the needs of society.

A new era of advanced, high-value world-class biomaterials, biomedical devices and implants is coming. Is India ready?

Appendix
Pedagogy and the Genesis of this Book

B. Basu, *Biomaterials Science and Implants*,
https://doi.org/10.1007/978-981-15-6918-0

Appendix A
Defining Key Elements of Biomaterials Science

The definitions in this section are reproduced from 'Definitions of Biomaterials for the Twenty-First Century', a review of key, critical biomaterial terms, and definitions endorsed by the International Union of Societies for Biomaterials Science and Engineering. The author attended this meeting and deliberated in the discussion with the global leaders in the field, leading to the finalisation of the following definitions.

Bioactivity A phenomenon by which a biomaterial elicits or modulates biological activity.

Bioceramic Any ceramic, glass, or glass–ceramic that is used as a biomaterial.

Biocompatibility The ability of a material to perform with an appropriate host response in a specific application.

Bioelectronic Implant Implants that could send an electrical pulse to a major nerve to alter the commands an organ receives, and thereby control its function.

Biofabrication The production of complex living and non-living biological products from raw materials such as living cells, molecules, extracellular matrices, and biomaterials.

Bioink Cell-laden biomaterial that can be integrated into additive manufacturing processes to print 2D and 3D structures.

Biomaterial Material designed to take a form which can direct, through interactions with living systems, the course of any therapeutic or diagnostic procedure.

Biomaterials Genome Integration of computational tools, large databases, and experimental techniques to explore the basic material elements and combine them to discover and design new biomaterials for medical products.

Biopolymer A polymer synthesised by living organisms.

B. Basu, *Biomaterials Science and Implants*,
https://doi.org/10.1007/978-981-15-6918-0

Bioprinting The use of 3D additive manufacturing technology that typically incorporates one or more biological components, viable cells, growth factors, and/or bioactive materials.

Biosensor A device used to detect the presence or concentration of a biological analyte.

Blood Compatibility/Haemocompatibility The ability of a blood-contacting biomaterial to (1) avoid the formation of a thrombus by minimal activation of platelets and of blood coagulation, (2) minimise activation of the complement system, and (3) minimise haemolysis.

Cell Therapy The process of introducing new cells into a tissue in order to treat disease.

Controlled Release Release of a solute, drug, diagnostic, or therapeutic agent from a carrier, system, or device in a planned predictable manner.

Drug Delivery Delivery and/or release of therapeutic and/or diagnostic agents.

Drug Eluting A device which (slowly) releases a drug to prevent cell proliferation.

Extracellular Matrix The composite material (ECM) in and between cells with both structural and regulatory function. ECM is composed of collagens, fibronectin, elastin, fibrillins, enzymes, proteoglycans, glycosaminoglycans, and others.

Gene Therapy The transplantation of normal genes into cells in place of missing or defective ones for the treatment of genetic disorders.

Graft Piece of viable tissue or collection of viable cells transferred from a donor site to a recipient site for the purpose of reconstruction of the recipient site.

Growth Factors Proteins that stimulate the activity of genes required for cell growth and cell division and may also mediate cellular migration, differentiation, and synthetic activities.

Hydrogel A physically or chemically cross-linked polymer, swollen in water or biological fluids.

Immunotherapy A treatment which elicits or suppresses an immune response.

Implant A medical device made from one or more biomaterials that is intentionally placed, either totally or partially, within the body.

Inflammation The response to injury and/or contact with biomaterials, involving a cascade of activation with blood cells and humoral factors, and both acute and chronic inflammation. In the case of implanted devices, this may lead to resolution with fibrous capsule formation and remodelling; with *ex vivo* devices, this may lead to systemic effects.

Mechanotransduction The processes by which cells sense mechanical stimuli and convert them to biochemical signals that elicit specific cellular responses.

Microfluidics The behaviour, control, and manipulation of fluids that are geometrically constrained to small, typically submillimeter, scale.

Nanomaterial Material with one or more external dimensions, or an internal structure, which could exhibit novel characteristics compared to the same material without nanoscale features.

Nanoscale Having one or more dimensions of the order of 100 nm or less.

Nanoparticle Particle with one or more dimensions at the nanoscale.

Nanotechnology The science of manipulating materials at the submicron level.

Organoid Self-organised three-dimensional tissues are typically derived from stem cells and embedded in a matrix, and mimic key functional, structural, and biological complexity of an organ.

Osseointegration The capability of substrate-adherent osteoblasts to produce bone.

Osteoclast Large multi-nuclear cell associated with absorption and removal of bone.

Osteoinduction Action or process of stimulating osteogenesis.

Prosthesis Device that replaces or supports a body part.

Remodelling The constant process of ECM and fibrous capsule collagen degradation and synthesis initiated and controlled by chemical, physical, mechanical, and other properties of the biomaterial in the *in vivo* environment.

Regeneration Synthesis, renewal, or growth of new functional tissue which has been lost due to injury, congenital deficiency or has failed due to ageing or disease.

Regenerative Engineering The integration of advanced materials sciences, stem cell science, engineering, developmental biology, and clinical translation for the regeneration of complex tissue and organ systems.

Regenerative Medicine Therapies that treat disease, congenital condition, and injury by the regeneration of functional tissue or organ structures.

Scaffold A biomaterial structure which serves as a substrate and guide for tissue repair and regeneration.

Self-assembly The autonomous organisation of components into patterns or structures without human intervention.

Stent A tubular support placed in a blood vessel, canal, or duct to aid healing, prevent or relieve an obstruction or stenosis.

Surface Topography Delineation of the natural and artificial features of a surface from a flat plane.

Targeting A process of drug or therapeutic agent delivery only to the desired site of action.

Template Biomaterials-based construct of defined size, chemistry, and architecture that controls the delivery of molecular and mechanical signals to target cells in tissue engineering processes.

Therapeutic Agent A substance used to treat a disease or other medical condition.

Tissue Engineering The use of cells, biomaterials, and suitable molecular or physical factors, alone or in combination, to repair or replace tissue to improve clinical outcomes.

Transplant Tissue structure, such as a complete or partial organ, that is transferred from a site in a donor to a recipient for the purpose of reconstruction of the recipient site.

Whole Organ Engineering The engineering of replacement organs that results in the preservation of tissue-type-specific matrix in a 3D architecture that closely mimics the native tissue and seeks to provide organ-specific functionality.

For e-learning resources, please see the National Programme on Technology Enhanced Learning (NPTEL) online courses, offered by the author;

1. Introduction to Biomaterials (https://nptel.ac.in/courses/113104009/)
2. Biomaterials for Bone Tissue Engineering Applications (https://www.youtube.com/playlist?list=PLLj8usx87A_b16gYoCJclV9ADCgnre0tq)

Appendix B
Ethical and Safety Considerations

Good laboratory practice (GLP) provides a framework of principles and standard operating procedures that establishes an ethically acceptable framework for any *in vitro*/*in vivo* study to produce reliable and high-quality results/outcomes. These studies should be planned, performed, monitored, reported, and archived according to GLP. GLP therefore assures regulatory authorities that the data in a report comply with the standard guidelines.

The ethical way to procure standard stem cell lines is from the stem cell banks that are available worldwide. These places store the cell lines under biologically appropriate conditions, which are further used for regenerative medicine and therapy.

Documentation for Ethical Clearance

The following are the documentations for ethical clearance that one is recommended to comply with:

- Proof of institutional ethical and legal approvals
- Donor informed consent documents
- Confidentiality on the information of donor
- All scientific details, like culture conditions, passage number, etc.

The investigator should read the ICMR-DBT 'Guidelines for Stem Cell Research and Therapy'. The documentation requires stating the kind of stem cells being used and its culture protocol in detail with information of the name of reagents being used, media composition, recovery and expansion of stem cells from frozen stock, their subculturing methods, and cryopreservation technique.

Every country has a different set of stem cell guidelines. The Indian Council of Medical Research (ICMR) in India last released the revised version of 'Ethical Guidelines for Biomedical Research on Human subjects' in 2017. This document enlists requirements and conditions for carrying out 'stem cell research and therapy'

B. Basu, *Biomaterials Science and Implants*,
https://doi.org/10.1007/978-981-15-6918-0

(SCRT). The investigators of a laboratory should ensure that the cell lines are maintained and used according to the guidelines of the document, as mentioned in the Material Transfer Agreement (MTA). While using stem cells, an investigator should declare the level of manipulation, that a given stem cell line is expected to experience during culture on biomaterials.

Recommended Guidelines for Biocompatibility Evaluation

Techniques and expertise from multiple disciplines, including biological and medical sciences, need to be combined for pursuing new research programmes in an interdisciplinary field, like biomaterials science. Clearly, there is a high need to incorporate ethical guidelines for conducting safe research in this field. This challenge deals with the fundamental principles and standard culture protocols for different cell types, including stem cells, preclinical studies, and clinical trials. This section includes the ethical guidelines for *in vitro* assays in a laboratory, and how to maintain a cell culture facility within a safe physiological environment. Additionally, biological assays that quantify cell viability, toxicity, differentiation, and cell fate processes are discussed in detail. This section outlines the ethics, and laboratory safety rules that are defined for conducting both *in vitro* and *in vivo* experiments with biomaterials.

No.	Recommended guidelines for biocompatibility evaluation
1.	Approval from the biosafety committee of the institution for setting up a cell culture facility in a laboratory and for conducting *in vitro* cell culture-related experiments. In addition, institutional stem cell committee's approval to conduct any study using stem cells
2.	The cells at their optimal growth stage need to be certified as healthy before testing the biomaterial on them. It is recommended to use $n < 5$ (n = passage number)
3.	For cell culture studies, the researcher should grow the cells in 3–4 times interval, depending upon the cells' doubling time. The MTT/LDH assays will be used to quantify cell viability, while fluorescence microscopy/scanning electron microscopy should be used for recording cell morphological changes. For polymeric or carbonaceous biomaterials, it is recommended to use the LDH assay
4.	For the biochemical assays, the reagents should not interfere or react with the biomaterial substrate constituents, in order to avoid errors
5.	The systematic increase in the cell numbers indicates positive compatibility of the biomaterial substrate with the cell line, and therefore culture experiments are to be conducted over 3–4 timepoints, depending on doubling time
6.	More than one cell type should ideally be used for testing the application of the biomaterial. For example, for bone replacement, fibroblast and osteoblasts can be used; while for neural applications, neuroblastoma and Schwann cells can be used
7.	Both positive and negative controls, along with benchmark controls, are highly recommended for biocompatibility assessment
8.	Approval from the human ethics committee of the institution for conducting haemocompatibility studies that require usage of human tissue cells or human blood

(continued)

(continued)

No.	Recommended guidelines for biocompatibility evaluation
9.	For blood compatibility assay studies, ISO guidelines need to be followed. This requires soaking the biomaterial at 37 °C for 30 min followed by recording the blood parameters
10.	Statistical software programs (SPSS, ANOVA, students' T-test, etc.) are to be used to estimate significance should be used while analysing the cytocompatibility/haemocompatibility data in any research publication or scientific document. A biostatistician can also be consulted in such cases

Appendix C
International Research Groups

A summary of the research activities in different groups around the world is presented below. This list, however, does not contain all the key international research groups, and many of these have carried out seminal work in the field of Biomaterials.

C.1 North America

Rajendra Bordia, Clemson University, South Carolina, USA His team investigates porous biomaterials for bone and joint implants. Stress shielding in solid metallic load-bearing bone replacement implants (e.g. hip stems) arises from a stiffness mismatch between the higher stiffness implant and the lower stiffness bone. This is believed to contribute to implant loosening and the need for a highly undesirable revision surgery. Bordia's group has used elective electron beam melting, to create engineered porous structures with controlled porosity and pore size in Ti-6Al-4V alloy. The group was the first to investigate the fatigue properties of these materials.

David Kaplan, Tufts University, USA The lab focuses on biopolymer engineering to understand structure-function relationships, with an emphasis on studies related to self-assembly, biomaterials engineering, tissue engineering and regenerative medicine. In addition, the lab has pioneered the study of silk-based biomaterials in regenerative medicine. His past contributions have resulted in the emergence of silk as a new option with excellent biocompatibility in the degradable polymer field, and new tissue-specific outcomes with silk as scaffolding in gel, fibre, film or sponge formats, with complex 3D tissue co-culture systems.

Cato Laurencin, University of Connecticut, USA The world-renowned group is working on regenerative engineering (a field Laurencin has founded), nanotechnology, polymeric materials science, bioreactor technology, drug delivery systems, morphogenesis, and stem cell technologies. The group, which was the first to publish on nano-fibre technologies for biomedical purposes, and the first to develop clinical

B. Basu, *Biomaterials Science and Implants*,
https://doi.org/10.1007/978-981-15-6918-0

polymer ceramic systems for bone regeneration, is now working on the development of whole limb regeneration using new paradigms in regenerative engineering.

Nicholas Peppas, University of Texas at Austin, USA He has provided new insights into biomaterials design and development, drug delivery systems and advanced, intelligent, feedback controlled biological systems. The group is known for its work on compatible, cross-linked polymers (hydrogels), which have been used as controlled release devices, especially drug delivery, peptides and proteins, development of novel biomaterials, biomedical transport phenomena, and biointerfacial problems.

John A. Rogers, Northwestern University, USA He leads a vibrant research group that seeks to understand and exploit interesting hybrid combinations of polymers, liquid crystals, and biological tissues, to induce novel electronic and photonic responses in these. Current research focuses on soft materials for conformal electronics, nanophotonic structures, microfluidic devices and microelectromechanical systems, with bio-inspired and bio-integrated technologies.

Molly Shoichet, University of Toronto, Canada She leads a group in researching biomaterials for regenerative medicine, tissue engineering and drug delivery. Her research program focuses on strategies to promote tissue repair after traumatic spinal cord injury, stroke and blindness and enhance both tumor targeting through innovative strategies and drug screening via 3D cell culture with new hydrogel design strategies.

C.2 Europe

Serena Best, University of Cambridge, UK The group has developed a range of synthetic substituted hydroxyapatite (HA) materials with physiologically relevant ionic lattice substitutions, for skeletal defect-filling and as scaffolds for tissue engineering. Her group is also working on bioactive and bioresorbable composites for tissue engineering, including biodegradable polymers, and bioactive ceramics, glasses and glass ceramics as fillers. One of her current research themes is to develop collagen-GAG based scaffolds for dental restorations, heart tissue repair and the production of platelets.

Aldo Boccaccini, Institute of Biomaterials, Erlangen, Germany His research group is developing nanostructured biomaterials for medical implants, tissue engineering of bone, cartilage and cardiac tissue and drug delivery. In bone tissue engineering, Aldo's group is focused on controlling surface topography and chemistry of 3D tissue engineered scaffolds, to enhance growth of new bone tissues and vascularisation. For cardiac tissue engineering, nanostructured polymer biomaterials are being developed as cardiac patches.

Marc Bohner, RMS Foundation, Switzerland Their research focus has been biomaterials, in particular bone graft substitutes and calcium phosphates. This comprises synthesis, nanostructuration and use of various calcium phosphates, the improvement of delivery techniques for bone substitutes and application of these materials, methods, and concepts in clinical use. The RMS Foundation is a research and testing institute offering testing services in the fields of medical and materials technology as well as mechanical process engineering.

Sarah Cartmell, University of Manchester, UK Her group is researching translation of novel tendon repair products, mechanotransduction in human mesenchymal stem cell differentiation, design of a novel co-culture bioreactor for osteochondral tissue-engineered plugs, use of statins for influencing osteoblast proliferation and differentiation, electrical stimulating bioreactors for modulating stem cell fate process and ligament tissue engineering.

Jérôme Chevalier, Institut National des Sciences Appliquées (INSA), Lyon, France His group is known for pioneering research on low temperature degradation of ZrO_2 bioceramics and developing new ZrO_2-based biocomposites for musculoskeletal applications, as well as for conducting many performance-limiting tests (e.g. Hip simulator test).

Christophe Drouet, CIRIMAT, Toulouse, France His team extensively uses biomimetic nanocrystalline apatites for production of highly bioactive bone scaffolds towards activation of existing biomaterials with low bioactivity. The high resemblance to bone mineral allows envisioning original applications beyond bone regeneration—increasingly appealing in view of ever-more challenging applications in medicine, whether for bone tissue engineering or nano-medicine.

Michael Gelinsky, TU Dresden, Germany His lab is working on biomaterials design for hard and soft tissues, scaffold-based tissue engineering and translational musculoskeletal medicine. Selected ongoing projects are material development for 3D printing and bioprinting applications, therapeutic drug/ protein delivery biomimetic bone matrices and biomaterials for articular cartilage defect healing.

Julian Jones, Imperial College London, UK His group is conducting research on strong porous scaffolds for regenerative medicine, porous materials for minimally invasive diagnosis, tough sol-gel hybrids, therapeutic nanoparticles, processing of glasses, bioactive materials, sol-gel chemistry and cell responses to biomaterials. Jones' research group has done seminal works on bioglasses, including nanoparticles, that can deliver therapeutic ions inside cells and "bouncy bioglass," sol-gel hybrid materials that can take cyclic loads, with mechanical properties that can be matched to that of host tissue.

Anne Leriche, Université Polytechnique Hauts-de-France, France Concerning bone substitute fabrication, she has a great expertise in calcium phosphate ceramic powder synthesis by coprecipitation method and in macroporous ceramics processing by three different methods. Current research concerns the impact of the ceramic part surface modified by laser patterning on cell colonisation. The second interesting expertise of the lab is the microwave assisted sintering, a new and original sintering technology allowing densifying ceramics faster than conventional heating techniques while maintaining equivalent and sometimes higher mechanical properties. This technology has successfully been applied to CaP macroporous scaffolds with considerable sintering time duration reduction. Thirdly, the lab is working on the synthesis and purification of polypeptides which exhibit osteoinductive and/or antibacterial properties using molecular biology techniques. The synthesised polypeptides are composed of dual protein, combining a collagen-like structure and an antibacterial peptide of human sequence to limit the immunogenicity risks. The polypeptides are grafted onto dense or porous ceramic parts to promote more rapid integration of the implant and to reduce bacterial infections post-surgery.

Ivan Martin, University Hospital of Basel, Switzerland His laboratory develops three dimensional *in vitro* model systems to explore fundamental facets of tissue development. The engineered nasal cartilage grafts developed in his laboratory have been successfully implemented in a phase I human clinical trial for nasal reconstruction after tumor resection. Another conclusive phase I clinical study using similar grafts for the regeneration of articular cartilage in the knee after traumatic injury has enabled to progress to the ongoing multicenter phase II human clinical trials to treat knee cartilage injuries. The applicability of engineered and decellularised tissues as templates for bone regeneration has been successfully tested and published by his team in high impact factor journals. For example, his laboratory is amongst one of the few tissue engineering laboratories that combine the principles of developmental biology to develop bone and cartilage tissue simulating events of embryonic development. Furthermore, his team has developed various bioengineering platforms like the bioreactor systems for the direct perfusion of culture medium and cellular suspensions through the engineered 3D constructs, and some are commercialised.

Abhay Pandit, National University of Ireland, Galway His research group has developed functional, biocompatible building blocks for targeted, controlled, drug-release and delivery systems. In addition, micro- and nano-structured implants and devices designed to emulate fundamental cellular architecture have facilitated clinical translation of cell-based devices.

Jouni Partanen, Aalto University, Finland The group conducts research in the field of Additive Manufacturing (AM). The emphasis has been in industrial applications of AM of the technology, including the potential new business models this digitally driven production method makes possible. The group also develops new variations of AM and other digitally driven production methods. In the medical field, the group has been involved in research and development to generate custom implants and medical products from person-specific scanned medical data.

Rui Reis, University of Minho, Portugal He is leading a vibrant research group on 3B (Biomaterials, Biodegradables and Biomimetics). Major research areas at his group include new materials development, drug delivery, tissue engineering, regenerative medicine, nanomedicine, and stem cell isolation and differentiation. Their goal is to develop novel biomaterials based on natural polymers for applications in drug delivery and tissue engineering of bone, cartilage and skin.

Jukka Seppälä, Aalto University, Finland His research combines the frontline biopolymer research with fast-developing digitisation, additive manufacturing and biomedical sciences. His research group is one of the global leaders in the biopolymer structure property correlation research. His research has led to groundbreaking results in polymer-based controlled therapeutic agents release systems, extremely strong orthopaedic fixation devices and additive manufactured scaffolds for tissue regenerative purposes. The key collaboration partner in the development of biomedical applications for nerve and bone regeneration has been Prof. Ashok Kumar of IIT Kanpur and his group. This collaboration has led to breakthroughs in nerve regeneration through 3D-printed and cryogel filled nerve conduits.

Anna Tampieri, Istituto di Scienza e Tecnologia dei Materiali Ceramici (ISTEC), Italy She leads the Bioceramics Research Group, focusing on the synthesis of bioinert and bioactive ceramics, like HA, TCP and non-stoichiometric apatites. In particular, her group aims to develop porous ceramics for bone tissue replacement and modulated drug delivery. Tampieri's group also works on biologically inspired CaP–based composites, such as HA/collagen.

C.3 Asia-Pacific

Dietmar W. Hutmacher, Queensland University of Technology, Australia The group has three main areas of research: cartilage, bone graft and 3D cell cultures. In the area of cartilage, the group's work on the molecular characteristics of zonal chondrocytes under dynamic cell culture conditions can lead to the development of a novel cartilage engineering technology platform that will allow for treatment of elderly patients and larger defects. Regarding bone graft, the group hypothesise that a composite scaffold (already successfully utilised in low-load bearing bone defects) can be biomechanically optimised and be combined with controlled delivery of angiogenic (PDGF/VEGF) and osteoinductive (BMP) molecules producing biologically active engineered bone graft systems with mechanical properties suitable for load-bearing applications.

Laura Poole-Warren, University of New South Wales, Australia She leads a research group in biomedical engineering focussed on design and development of novel materials for neural electrodes and other soft-tissue interfaces, and on 3D engineered neural tissue models. A key theme of her work is on design of new biosynthetic materials and understanding material and device interactions with cells and tissues. Her group also studies hydrogel drug delivery, biomaterials for wound healing, tissue engineering percutaneous access catheters and vascular grafts.

Yarlagadda Prasad, Queensland University of Technology, Australia His lab conducts research on nanopillar surfaces and surface modifications with bactericidal effect on titanium implants, additive biomanufacturing and bending of fracture fixation plates in orthopaedic surgery. Also, his research group is investigating antiviral and antibacterial nanostructured surfaces for hospital applications.

Seeram Ramakrishna, National University of Singapore His lab is working on electrospinning and nanostructured materials. He demonstrated that nano-fibres are the ideal scaffolds to provide biomimetic microenvironment for different cells in tissue engineering application. His team is now investigating the use of their patented nanoyarn and 3-D nanofibrous scaffold in stem-cell capture and expansion. Seeram's team successfully used nano-fibres with metal oxides and organic molecules to detect and neutralise chemical warfare agents, such as, nerve gas and mustard gas, more effectively than the current systems.

Appendix D
Conferences and Workshops

This section includes summaries of events, wherein extensive discussions took place over the last three years. Many such discussions allow the author to gain an understanding of the status and challenges in the field of biomaterials and implants in India. There are several technical details that were discussed which are not summarised here, since they are out of the scope of this monograph.

Clinical-Focused Workshop: New Materials for Healthcare: Idea Generation Workshop at Hotel Oberoi, Bangalore, 6 May 2018

The Indian Institute of Science (Bangalore) and Tata Steel New Materials Business hosted a unique conclave, 'New Materials for Healthcare' at Hotel Oberoi, Bangalore, on 6 May 2018. The event was endorsed by the American Ceramic Society. The central objective of this conclave was to understand the challenges faced by the active clinicians, which remains to be addressed with translational research involving innovative biomaterials. The conclave brought together 12 clinicians in the fields of orthopaedics, dentistry, neurosurgery, ear, nose and throat (ENT), and urology. There were seven representatives from six industries, nine young faculty colleagues from IITs and IISc, and nine young researchers. There were five focused sessions in the conclave which were moderated by academic researchers. Each session started with a clinician's presentation, followed by presentations of academic researchers and industry representatives. Each session ended with a panel discussion on the session theme.

1. **Orthopaedics**

In this session, the pros and cons, mechanical properties and issues of cost, and quality of titanium and stainless steel implants were discussed. The problem of Ti-based screws getting stuck in the implant during fracture fixing surgeries was

B. Basu, *Biomaterials Science and Implants*,
https://doi.org/10.1007/978-981-15-6918-0

discussed. Potential solutions are the usage of PEEK polymeric screws and the usage of additive manufacturing as a technique for developing patient-specific implants and accessories. Recent advances in the use of a biodegradable polymeric scaffold in orthopaedic applications for osseointegration and reunion of fractures were discussed.

The problem of osteoarthritis, which is associated with chronic joint pain, was discussed. The potential solution is an implant with improved wear resistance and mechanical strength. Development of a chondrogenic scaffold and injectable cartilage are potential solutions to osteoarthritis. Osteomyelitis is associated with infection at the fracture site of bone which might spread to the nearby tissue. There exists a need for an affordable solution. Biocomposite mixed with antibiotics (STIMULAN®) is already in the Indian market, but the price is still not easily affordable.

2. **Neurosurgery**

During this session, the design of materials that improve interaction between neurons, bone, and meninges was discussed. Also, prevention of the formation of scar tissue by astrocytes is extremely important for neurological/cranial-based applications. Three main applications were highlighted: childhood skull deformation, adult skull deformation, and duraplasty. For the first two applications, the problems associated with titanium implants were discussed. For duraplasty, the pros and cons of polypropylene- and collagen-based implants were discussed. For the craniectomy application, titanium- and stainless steel-based alloys have been established as the most promising implant materials. However, cost is a major limiting factor. Biodegradable polymers are most widely used in cranial applications, as screws and cranial plates; however, breakage during implantation is a disadvantage.

3. **Urology**

Malignancy in the bladder, urethra, and kidneys are major clinical applications. There is a current need for a new material for replacing the bladder in bladder transplantation. The neo-bladder should have the right mechanical properties and should be encrustation resistant. For surgery related to the urethra, doctors require a biodegradable material for stents. During kidney surgery, there is need for a haemostatic agent which can stop blood flow temporarily without a need to clamp the blood vessels. There is need for an affordable reusable organ retrieving bag made from a new polymeric material, which should stay intact during the surgical procedure. For incontinence, there is need for a device which can manage the flow of urine into a urinary bag.

4. **ENT: Cochlear implants**

Insertion trauma and long-term inflammatory damage to hair cells leading to loss of residual hearing is a key application. Cochlear implants are one option for treatment of hearing loss. Development of an affordable indigenous implant is preferred since the cost of imported implants is high. Development of an electrode material with improved biocompatibility is needed.

5. **Dental**

Three areas for potential innovation were discussed. Abfraction is a physico-mechanical loss of tooth architecture that is not caused by tooth decay, located along the gum line. This affects mostly the enamel and the dentin part of the crown. Development of moderate and steady peptide-releasing biomaterials which will be able to regenerate or cure the area as well as a temporary supporting material in the affected region may be a probable solution. Affordable digital impressioning is to be developed which will allow the surgeon to reconstruct tooth structures using CAD/CAM technology only by relying on the digital scan of the neighbouring dental architecture. Caries can be described as cracks, channels, pits, grooves, and cavities in tooth structure which can originate in the back teeth, between teeth, around dental filling or bridgework, and/or near the gingival margin. Development of a suitable biomaterial which will be able to form a strong and long-term bond with dentin for treatment of caries can be a potential solution.

Biomaterialomics—Focused Workshop: ICME Approaches to Innovation in Biomedical Implants Workshop at IISc, Bangalore, 10–12 August 2018

The Indian Institute of Science (Bangalore) and ICME National Hub@IIT Kanpur hosted the National workshop 'ICME Approaches to Innovation in Biomedical Implants' from 11 to 12 August 2018 along with a pre-workshop tutorial on 10 August 2018 at the Materials Research Center, Indian Institute of Science, Bangalore. This event was additionally supported by the DBT Bioengineering and Biodesign Initiative and the DBT Centre of Excellence on Biomaterials, and was organised as one of the activities of the DST-SERB-funded Vajra scheme to Professor Surya R. Kalidindi from Georgia Institute of Technology and Professor Bikramjit Basu from Indian Institute of Science (Bangalore). The workshop was endorsed by the American Ceramic Society, the Biomedical Engineering Society and the Society of Biomaterials and Artificial Organs.

In the pre-workshop tutorial aimed at introducing cutting-edge research topics to young researchers (i.e. graduate students and post-doctoral students), Prof. Kalidindi delivered a series of lectures on the foundational elements of the emergent Integrated Computational Materials Engineering (ICME) and Materials Genome Initiative (MGI), whose goal is to accelerate dramatically the rate at which new and improved materials (including biomaterials) are designed, developed, and deployed in biomedical applications. Additionally, a number of experts from complementary fields of expertise of central focus to the workshop (including biomaterials, bioimplants, and bioengineering) also presented lectures in the pre-workshop tutorials. The topics covered in the pre-workshop included statistical quantification and low-dimensional representation of material structure, techniques for the measurement of microstructures over multiple length scales, accelerated development of high-performance materials using machine learning, formulation of process-structure-property (PSP) linkages using machine learning tools, phase-field modelling in ICME, molecular dynamics simulations in modelling of biomaterials, high-throughput experimental assays for PSP linkages, quantitative biology of the cytoskeleton, immune response to biomedical implants, and tissue engineering applications in cardiovascular surgery.

The main workshop explored the cutting-edge research opportunities and potential benefits from the application of ICME-based approaches to the design and development of implantable biomaterials. In the field of materials science, new protocols are being developed for the extraction of reliable and robust PSP linkages from all available experimental and simulation data sets. It is anticipated that the formal application of these techniques to implantable biomaterials will lead to accelerated and rational development of biomedical implants and devices. The lectures and discussions in this workshop deliberated on topics such as clinical challenges in the successful deployment of biomedical implants, indigenous manufacture of bioimplants, and the multi-scale physics-based simulations for integrated materials, product design, and development. A common thread in all the discussion was the new opportunities afforded by the emergent tools in data science and informatics for circumventing many of the current challenges in the biomaterials/bioengineering fields. In particular, there was excitement that these new tools can help address the current major gaps in our fundamental understanding of how the microstructure and material composition influence cell functionality, bone remodelling, genotoxicity, and osseointegration. It was broadly understood that a physics-based understanding, capturing, and exploiting of the principles of biocompatibility in a consistent PSP framework would allow for a rational design of bioimplants. It is expected that future research at the intersection of biomaterials, bioimplants, materials science, manufacturing, data sciences, and informatics holds tremendous promise for arriving at new and improved materials at dramatically reduced cost and effort.

The workshop brought together 35 speakers, which include academicians, clinicians, and industrialists. There were 5 representatives from 4 industries, 5 clinicians, 25 faculty colleagues from IITs, NITs, and IISc, and 25 young researchers. There were nine sessions chaired by different experts in the field. There was also a panel discussion with lively discussion among all of the attendees. The panelists included leaders from academia, government, and industry.

National Status and Road-mapping Workshop: Fundamentals of and Clinical Perspectives on Implantable Biomaterials 23–24 August 2019

A one and a half day workshop was held from 23 to 24 August 2019, at Indian Institute of Science, Bangalore, featuring Prof. Sarah Cartmell, from the University of Manchester, foreign Principal Investigator under the Scheme for Promotion of Academic and Research Collaboration (SPARC). The event was hosted by Prof. Bikramjit Basu, Principal Investigator of the project. The presentation sessions explored ongoing research in the field of biomaterials science, tissue regeneration, and bioengineering, India, and were open to students of Indian Institute of Science. Faculty presenting were from IIT (BHU), Varanasi, NITK Surathkal, IIT Kharagpur and IIT Roorkee.

The workshop featured one-to-one discussions of faculty with Prof. Cartmell. Prof. Basu discussed challenges and recommendations associated with the current research ecosystem, in preparation for publishing a monograph for the Indian National Science Academy on the subject. How to map the terrain ahead for improving research on indigenous biomedical implants and the related ecosystem were deliberated on.

International Conferences and Symposia

International Symposium 'Fundamentals of and Clinical Perspectives on Biophysical Stimulation of Cells on Implantable Biomaterials'—GFMAT-2 and Bio-4 at Toronto, Canada, 21–26 July 2019

In the interdisciplinary field of biomaterials, the phenomenological interaction of a biological cell on a material substrate under normal culture conditions is broadly known, and researchers use many approaches to tailor substrate modulus or surface wettability in an effort to enhance cell–material interaction. In this context, a strikingly different approach in the field of biomaterials science has recently been proposed to replicate the dynamic physiological micro-environment of native tissues. Using experiments and modelling approaches, the recent research has witnessed a new dimension to the design of biomaterials by developing an integrated approach towards the development of new culture platforms and to carry out thought-provoking experiments to generate a fundamental understanding on the external field-mediated cellular functions. To discuss recent advances, a special symposium was organised by Bikramjit Basu. This symposium was attended by experts on biomaterial development, biology, medicine. Effective discussions on the research collaborations were also important subjects for future international projects. A forum such as this conference is expected to accelerate progress of the biophysical stimulation approach for human healthcare applications.

International Conference on the Design of Biomaterials (BIND 12) and the Young Scientist Training Workshop (BIOMAT 12) at Indian Institute of Science, Bangalore, 9–12 December 2012

The global biomaterials market is estimated to be worth $88.4 billion USD by 2017. A rise in musculoskeletal disorders further enlarges the scope of biomaterials in healthcare, which impose a burden of >$250 billion USD every year on society. It is interesting to note that North America has the largest biomaterials market share in the world. However, the Indian market is slowly emerging.

In lieu with the emerging trends, a three-day international conference on design of biomaterials (BIND 12) was organised at Indian Institute of Science, Bangalore in December last year to (i) identify problems associated with biomaterials design and fabrication, (ii) review the progress being made in this important area in India vis-à-vis the rest of the world, (iii) discuss the development of novel techniques integrating nanotechnology, stem cell, or regenerative medicine and tissue engineering and (iv) to serve as a platform for young researchers to present their results before the biomaterials community.

BIOMAT-12, a workshop preceded the conference, which was inaugurated by K. B. R. Varma (Chairman, Materials Research Centre, IISc). The main objective of the pre-conference workshop was to educate and train the next generation of Indian biomaterial researchers, and to interface with the Department of Science & Technology and Defense Research and Development Organization. The workshop was designed to cover basic concepts such as structure, properties, processing, and characterisation of biomaterials, and the evaluation of blood, cell, and bacterial compatibility of materials, to advanced topics such as tissue scaffolds, drug delivery, and imaging. Lectures were supplemented with demonstrations and laboratory visits. Overall, the workshop served as a platform to disseminate information to future stakeholders on the subject through a series of classroom lectures and laboratory training sessions.

The design and development of biomaterials require integration of concepts and expertise of two widely different disciplines—materials science and engineering and biological science. In order to ensure the continuous growth of this emerging field, numerous initiatives have been taken at the Indian Institute of Science, Bangalore. Also, several academic courses are being taught at the undergraduate and graduate levels in top most universities, including the IITs. M. S. Valiathan, the brain behind India's first artificial heart valve emphasised on the need for developing biomedical devices, and stressed that the events, like BIOMAT 12 workshop and BIND 12 conference, would act as a catalyst to stimulate young minds towards creating affordable healthcare in India. P. Balaram (former Director, IISc) stated the growing need for interdisciplinary research, education, and introduction of recent interdisciplinary programmes at the interface of engineering and biological sciences.

Different aspects of research related to the development of orthopaedic biomaterials, cell–material interactions, novel biopolymers, nanobiomaterials, and tissue engineering were discussed by 33 invited speakers and was attended by close to 200 delegates from USA, Belgium, Germany, UK, France, Portugal, Nepal, and various Indian institutes.

The advent of nanotechnology has made a significant impact on many areas of our lives, in particular, the area of electronics with faster and smaller devices flooding the consumer market every day. Nanotechnology promises to revolutionise healthcare through innovations in imaging, drug delivery, and 3D tissue scaffolds. A session on nanobiomaterials addressed challenges in making nanobiomaterials, their applications and toxicity, a prime concern.

Orthopaedics, one of the foremost branches of medicine, where the use of biomaterials has resulted in significant clinical success, has several issues related to prosthesis such as plates, screws, and joints primarily made of metallic alloys that are widely used. There is a need to improve the performance of these medical devices. Narendra B. Dahotre (University of North Texas, USA) discussed various surface engineering strategies to achieve appropriate surface morphology and chemistry. It is being increasingly realised by the biomaterials community that conventional manufacturing and materials processing techniques are inadequate to meet the specialised needs of biomedical application of materials. Thus, novel techniques are required to synthesise, process, and characterise biomaterials. Extent of osseointegration is enhanced by providing the desired pore size distribution and pore interconnectivity.

The challenges of working with stem cells and their interactions with biomaterials were taken into account. Given their pluripotency and self-renewal capacity, stem cells are likely to be the major cellular sources for different regenerative medical therapies.

International Conference on Design of Biomaterials (BIND-06) at IIT Kanpur, December 2006

One of the motivations to organise BIND-06 was to bring together the leading medical practitioners, biologists, and materials scientists from around the globe to share their excitements in the emerging area of biomaterials science. The conference was thus named due to its focus on discussing the frontier areas of new bulk/coating materials for biomedical applications. Such materials are synthetically produced outside the human body and once found suitable, are used in human implants. Therefore, they essentially integrate with the living organs and participate in the functioning of living systems. Another reason is that the research on biomaterials essentially '**binds**' biologists/biotechnologists with engineers (mechanical, chemical, metallurgy)/materials scientists and clinicians (orthopaedic surgeons, dental surgeons, ophthalmic surgeons, and cardiac surgeons). An appropriate binding of research ideas/expertise ensures a comprehensive understanding of the processing-microstructure-properties (physical as well as *in vitro*/*in vivo*) of biocompatible materials.

Among other directions, the conference focused on some topical areas, including bioceramics, polymeric biomaterials, cell–biomaterial interaction, and tissue engineering. The advances in the areas like tissue engineering, bioceramics, polymeric biomaterials, orthopaedic biomaterials, cardiovascular biomaterials, nanobiomaterials, ophthalmic biomaterials, dental biomaterials, and biomaterial applications were largely discussed in the symposium. Besides various technical sessions, a panel discussion on 'Biomaterials Education, Research and Industries: Future Perspectives', focused sessions to explore future possibilities of international collaboration with foreign delegates, were also organised during BIND-06. Two separate poster sessions were held for students/researchers and a total of 70 abstracts were judged by an expert panel of five judges (each session) for three best awards in each session. In addition, two parallel sessions (one on metals and ceramics and other on polymeric biomaterials and drug delivery), focused on students' oral presentation, were also held. These sessions were also judged by a panel of three experts in each session.

More than 250 delegates, including experts from 14 countries, namely Australia, Belgium, France, India, Japan, Malaysia, Nepal, Portugal, Singapore, Sweden, Turkey, UK, and USA, participated in BIND-06. In particular, eight invited speakers from USA delivered Plenary, Keynote and invited talks. The US participants were drawn from University of Virginia, Rensselaer Polytechnic Institute, Iowa State University, Columbia University, University of North Texas, University of Texas at San Antonio, NASA Glenn Research Center, New York University. In addition to 8 invited speakers from USA, the conference had a total of 19 invited speakers, of which 6 were from India and the remaining was from foreign countries, other than USA. The Indian participants were from institutions like IITs, SCTIMST, NIPER, Institute of Life Sciences, SASTRA, AIIMS, Reliance Life Sciences, Mumbai, Hindustan Latex, etc. In addition, top officials from government funding agencies (DST, DRDO, IUSSSTF) also participated in the conference.

Appendix E
Road-Mapping Meetings

First Meeting of Biomedical Implants Materials Mission (BIM^2) Mission at KIHT, 19 November 2018

The Kalam Institute of Health Technology (KIHT), Visakhapatnam, hosted a daylong partnership meet with key stakeholders of the medical implants industry to explore solutions to promote indigenous innovation in the critical health technologies. The meet was held in the premises of Andhra Pradesh MedTech Zone (AMTZ) and featured an august gathering of experts representing the biomaterial segment of medical technology. The meet was organised as part of the initiative of KIHT to promote healthcare innovation needs of the country, and the focus of the meet was on biological implantable materials, their sourcing, manufacturing, and clinical applicability in the Indian healthcare context.

Stakeholders represented by various reputed institutions like IITs, NITs, IISc, VIT, National labs, MSME manufacturers, large-scale manufacturers like Tata Steel New Materials Business, and policy facilitation units like KIHT and AMTZ, etc. participated. The meet presented a forum for active discussion on biomaterial implants innovations across the country including detailed perspectives of biological materials and the need to promote more indigenous innovation in the country which could lead to reduction in India's import dependency.

KIHT and AMTZ, being the forebearers of the medtech sector, took up the mandate to facilitate the requisite academic research projects needed to support manufacturing in a timebound frame and bring together stakeholders to create the first ever platform for end-to-end biomaterials product realisation. Speaking on the occasion, Dr. Jitendar Sharma, Managing Director AMTZ, called for closer cooperation among academic and industry in the biomaterials sector to address India's concerns on import dependency in implantable materials.

B. Basu, *Biomaterials Science and Implants*,
https://doi.org/10.1007/978-981-15-6918-0

Second Meeting of Biomedical Implants Materials Mission (BIM^2) at KIHT, 22 December 2018

The Kalam Institute of Health Technology (KIHT), Visakhapatnam, hosted a daylong annual review meeting of the Translational Centre of Excellence on Biomaterials for Orthopaedic and Dental Applications and also a second meeting of the Biomedical Implants and Materials Mission (BIM^2). This meeting reviewed and promoted indigenous inventions in the health technology centre across the country. This event was financially supported by the Department of Biotechnology, Government of India and the Ceramic and Glass Industry Foundation, American Ceramic Society. The objective of this meeting was to discuss healthcare innovations in orthopaedic and dental applications, in particular their related research, manufacturing and clinical applicability in the Indian healthcare context. This meeting emphasised the detailed intense knowledge on designs, applicability, compatibility with living systems, non-contamination attribute, antimicrobial sterilisation needs, etc.

Stakeholders from various reputed institutions like IITs, NITs, IISc, VIT, National labs, MSME manufacturers, large-scale manufacturers like Tata Steel New Materials Business, policy facilitation units like KIHT and Andhra Pradesh MedTech Zone (AMTZ), etc. participated in the meeting. Representatives from Faberz Technology Pvt. Ltd, Modern Bioceramics Pvt. Ltd and DUCOM Instruments Pvt. Ltd participated. Mentoring Committee members present were Prof. Abhay Pandit, University of Ireland, Galway, Mr. Ravi Sarangapani, Smith & Nephew, India, Prof. Surya

Kalidindi, Georgia Institute of Technology, Prof. Rajendra Bordia, Clemson University, Prof. H. S. Maiti, Indian Institute of Technology, Kharagpur and Dr. Jitendar Sharma, Kalam Institute of Health Technology.

This meeting started with opening remarks and a welcome address by Dr. Jitendar Sharma, Executive Director, MD, AMTZ. KIHT. Further, speaking on the occasion was Prof. Bikramjit Basu, Professor of the Materials Research Centre, Indian Institute of Science and Principal Investigator of the Translational Centre of Excellence. Prof. Basu is currently leading the Academic Consortium of the BIM^2 mission. He is also a Global Ambassador of the American Ceramic Society. Several Centre of Excellence Mentoring Committee Members and Co-investigators presented their research and experience in the field of biomedical materials research.

Third Meeting of the Biomedical Implants Materials Mission (BIM^2) Mission at IISc Bangalore, 26 February 2019

Representatives from TATA Steel New Materials Business, Kalam Institute of Health Technology, Intellectual Property and Technology Licensing Cell of IISc and Prof. Bikramjit Basu attended this meeting, where it was agreed that the key factors for biomedical products under BIM^2 are affordability in the Indian market, indigenous products and technologies, positive research outcomes, translatable technologies and acceptance by clinicians. The vision, mission, and objectives were further refined, as well as funding aspects, the model for the Consortium, running of projects and legal aspects.

Brainstorming Meeting at IIT Delhi and INSA. New Delhi, 26–27 August 2019

The author held a brainstorming meeting to discuss recommendations for this monograph with Prof. Neetu Singh, IIT Delhi, Prof. Shalini Gupta, IIT Delhi, Prof. A. B. Pandit and Prof. Naresh Bhatnagar, IIT Delhi, on 26–27 August 2019.

Centre of Excellence Meetings: Translational Centre on Biomaterials for Orthopaedic and Dental applications

Centre of Excellence Review Meeting at IISc, Bangalore, 11 August 2018

Representatives from IISc Bangalore, SCTIMST Thiruvananthapuram, DUCOM, ANTS Ceramics, KIHT, King George Medical University Lucknow, Georgia Institute of Technology, VNIT Nagpur, Datta Meghe Institute of Medical Sciences and TATA Steel New Materials Business attended. Prof. Basu emphasised the importance of developing a strong cohesive collaboration with KIHT and agreed to provide a list of innovations to KIHT with assessment of areas for collaboration. Prof. Basu commented on the market research for the femoral head and that the manufacturing cost for the product for the market is Rs 20 crores. Mr. Kingshuk Poddar suggested that the manufacturing panel of KIHT can assist with this product. Mr. Poddar commented that the involvement of industry while developing the prototype varies—whether it takes on a mentoring role and whether it has a stake during prototype development. He advised that industry should be involved from Day 1. He added that there are issues on both sides to be worked out, and that industry should be involved in meetings with academics so that both sides can come to a common ground. Dr. Zahir Quaziuddin said that once certification is given for a product, there should be no problem of acceptance by doctors.

Centre of Excellence Meeting at IISc, Bangalore, 25 July 2017

Representatives from Smith and Nephew met with Prof. Bikramjit Basu and researchers from IISc. Prof. Basu emphasised the need for creating a bridge between the work of academic researchers and commercial companies to solve the unmet needs in the medical field. Dr. Murphy, VP, New Product Development, Smith and Nephew, proposed to extend their cooperation in the case of solving medical challenges. Mr. Ravi Sarangapani, Director, New Product Development, Smith and Nephew mentioned that the biggest paradigm shift in this field of medical research would be the development of regenerative cartilage and that would allow surpassing of most challenges faced by the use of implants.

Centre of Excellence Second Review Meeting at IISc, Bangalore, 20 December 2016

In this meeting, there were representatives from SCTIMST Thiruvananthapuram, CGCRI Kolkata, IISc Bangalore, MSRDC Bangalore, IIEST Shibpur, and CIPET Chennai. Ms. Mala Srivastava from Nextvel Consulting LLP spoke on the regulatory policy status for Medical Devices in the backdrop of the US-FDA and that of Europe. It was suggested to follow universally accepted 'Standard Protocols' to achieve greater reliability and reproducibility of data. Credible correlation of data generated from different laboratories and institutes is essential. All projects under COE need to be executed in a GLP/GMP compliant manner. The need to understand the regulatory issues and requirements along the pathway of translation was emphasised.

Centre of Excellence Meeting on Biomaterials at IIEST, Kolkata, 18 December 2015

In this meeting, several aspects of the COE were refined, such as details of the Mentoring Committee, review meetings, and monitoring of projects. Prof. Bikramjit Basu, Prof. Abhay Pandit, Prof. Brian Derby, and IIEST researchers were present. It was decided that a special session on entrepreneurship will be organised alongside the review meetings. Also, MoU, Confidentiality, IP transfer among partner organisations are to be established. It was suggested to have an exchange of manpower among the participating institutes to facilitate training of young researchers.

Additional Reading and References

Agrawal, C. M., Ong, J. L., Appleford, M. R., & Mani, G. (2014). *Introduction to biomaterials: Basic theory with engineering applications.* Cambridge: Cambridge University Press.

Alberts, B., Johnson, A., Lewis, J., Raff, M., Roberts, K., & Walter, P. (2002). *Molecular biology of the cell.* New York: Garland Science.

Alijani, S., Anvari, A. (2018). Cycle numbers to failure for magnesium and its alloys in human body fluid. *Journal of Chemical Engineering and Materials Science, 9*(1) 1–8.

Alliance for Regenerative Medicine. Pharma & Biotech Survey, (2014). Accessed September 16, 2020. http://alliancerm.org/sites/default/files/ARM_Pharma_SurveyRept_Mar2014_e.pdf.

Allied Market Research. https://www.alliedmarketresearch.com/.

Allison, J., Backman, D., & Christodoulou, L. (2006). Integrated computational materials engineering: A new paradigm for the global materials profession. *JOM Journal of the Minerals, Metals and Materials Society, 58,* 25–27.

American Heart Association. (2008). Heart Disease and Stroke Statistics—2008 Update. *Circulation, 117*(4), 25–146.

Analysis of enzymatically amplified β-globin and HLA-DQα DNA with allele-specific oligonucleotide probes. *Nature, 324,* 163–166.

ARC Training Centre for Innovative Bioengineering, Australia. https://arctcibe.org/

Association of Indian Medical Device Industry. http://www.aimedindia.com/aboutus.html

ASTM WK65476. (2020). *Standard guide for characterizing fiber-based constructs for tissue engineered medical products.* Accessed September 16, 2020. https://www.astm.org/DATABASE.CART/WORKITEMS/WK65476.htm

ASTM WK62115. (2020). *Standard test method for measuring cell viability in a scaffold.* Accessed September 16, 2020, https://www.astm.org/DATABASE.CART/WORKITEMS/WK62115.htm.

Australian Safety and Efficacy Register of New Interventional Procedures—Surgical (ASERNIP-S) Group, Systematic review of needs for medical devices for ageing populations, World Health Organisation, 2015, Switzerland

Basu, B., Katti, D., & Ashok Kumar (2009). *Advanced biomaterials: Fundamentals, processing and applications.* USA: Wiley, and American Ceramic Society.

Basu, B., Keshavan, N., Bhagawati, B., & Bhattacharjee, D. (2019). *Biomedical implants and materials: Innovation, opportunities and challenges.* Kolkata: Think Pot Advertising Pvt. Ltd.

Biotechnology Industry Research Assistance Council (BIRAC). https://www.birac.nic.in/.

Brent, B., Farkas, D. L., Lansing Taylor, D., & Lanni, F. (1993). Enhancement of axial resolution in fluorescence microscopy by standing-wave excitation. *Nature, 366,* 44–48.

Balani, K., Verma, V., Agarwal, A., & Narayan, R. (2015). *Biosurfaces: A materials science and engineering perspective.* Wiley: Hoboken.

B. Basu, *Biomaterials Science and Implants*,
https://doi.org/10.1007/978-981-15-6918-0

Barnes Frank, S., & Greenebaum, B. (2007). *Biological and medical aspects of electromagnetic fields*. Boca Raton: CRC Press.

Barsoukov, E., & Macdonald, J. R. (2005). *Impedance spectroscopy: Theory, experiment, and applications*. Hoboken: Wiley.

Basu, B. (2017a). *Biomaterials for Musculoskeletal regeneration: Concepts*. Berlin: Springer Nature.

Basu, B. (2017b). *Biomaterials science and tissue engineering: Principles and methods*. Cambridge: Cambridge University Press.

Basu, B., & Balani, K. (2011). *Advanced structural ceramics*. Hoboken: Wiley, USA: American Ceramic Society.

Basu, B., & Ghosh, S. (2017). *Biomaterials for Musculoskeletal regeneration: Applications*. Berlin: Springer Nature.

Basu, B., & Kalin, M. (2011). *Tribology of ceramics and composites: Materials science perspective*. Hoboken: Wiley, USA: American Ceramic Society.

Biomaterials research in India: from tissue regeneration to drug targeting technology—India Bioscience. https://indiabioscience.org/columns/general-science/biomaterials-research-in-india-from-tissue-regeneration-to-drug-targeting-technology.

Biomedical engineering institutions in India. http://www.icehungary.net/download/fullp/full_papers/full_paper455.pdf.

Bennett, B. T., Hess, S. T., & Bewersdorf, J. (2008). 'Three-dimensional sub-100 nm resolution fluorescence microscopy of thick samples. *Nature, 5*, 527–529.

Brakenhoff, G. J., van der Voort, H. T. M., van Spronsen, E. A., Linnemans, W. A. M., & Nanninga, N. Three-dimensional chromatin distribution in neuroblastoma nuclei shown by confocal scanning laser microscopy. *Nature, 317*, 748–749.

3B's Research Group. https://3bs.uminho.pt/

Smith, B. C. (2011). *Fundamentals of fourier transform infrared spectroscopy*. CRC Press (Taylor & Francis).

Suzanne, C., Chang, S.-Y., Gravitt, P., & Respess, R. (1994). Long PCR. *Nature, 369*, 684–685.

Centre for Translational Bone, Joint and Soft Tissue Research. https://tu-dresden.de/med/mf/tfo/das-institut/profil?set_language=en.

Cleveland, W. S. (2001). Data science: An action plan for expanding the technical areas of the field of statistics. *ISI Rev, 69*, 21–26.

Cohen, S., & Leor, J. (2004). Rebuilding broken hearts. *Scientific American, 291*(5), 44–51.

Commission Matériaux Pour La Santé (MatSan). https://sf2m.fr/commissions-thematiques/commission-materiaux-pour-la-sante/.

Cúram Centre for Research in Medical Devices. http://www.curamdevices.ie/curam/.

Definitions of Biomaterials for the Twenty-First Century. (2018). *Proceedings of a Consensus Conference Held in Chengdu Province*, People's Republic of China.

Dental Implants Market Research Report. Mordor Intelligence. (2018). https://www.mordorintelligence.com/industry-reports/dental-implants-market.

Dental Implants: Global Markets to 2023. (2018). https://www.bccresearch.com/market-research/healthcare/dental-implants-global-markets.html.

Department of Biotechnology. http://dbtindia.gov.in/.

Dong-An, W., Varghese, S., Sharma, B., Strehin, I., Fermanian, S., Gorham, J., et al. (2007). Multifunctional chondroitin sulphate for cartilage tissue-biomaterial integration. *Nature Materials, 6*(5), 385.

Drew, S., Bao, G., Bettinger, C., Leong, K., Peshwa, M., & Ye, K. (2015). Global Assessment of Biological Engineering & Manufacturing, World Technology Evaluation Center (WTEC). https://haseloff.plantsci.cam.ac.uk/resources/SynBio_reports/BEM-FinalReport-Web.pdf.

Drosback, M. (2014). Materials genome initiative: Advances and initiatives. *JOM Journal of the Minerals, Metals and Materials Society, 66*(3).

EC Europa. *Biomaterials for Health: A Strategic Roadmap for Research and Innovation: Horizon 2020*. https://ec.europa.eu/research/industrial_technologies/pdf/biomaterials-roadmap-for-horizon-2020_en.pdf.

Engineering and Physical Sciences Research Council (EPSRC). https://epsrc.ukri.org/.

Free World Maps. https://www.freeworldmaps.net/powerpoint/.

Freshney, R. I. (2005). *Culture of animals cells*. Hoboken: Wiley.

Frost and Sullivan. (2020). *Market study of advanced ceramics and composites in the healthcare space*.

Fullwood, D. T., Niezgoda, S. R., Adams, B. L., & Kalidindi, S. R. (2010). Microstructure sensitive design for performance optimization. *Progress in Materials Science, 55*, 477–562.

Garcia, L., Soliman, S., Francis, M. P., Yaszemski, M. J., Doshi, J., Simon, Jr. C. G., & Robinson-Zeigler, R. (2020). Workshop on the characterization of fiber-based scaffolds: Challenges, progress, and future directions. *Journal of Biomedical Materials Research Part B: Applied Biomaterials, 108*(5), 2063–2072.

GlobalData Medical Intelligence Centre. https://www.globaldata.com/.

Govt plans to set up body to regulate medical devices sector—Business Standard. https://www.business-standard.com/article/economy-policy/govt-plans-to-set-up-body-to-regulate-medical-devices-sector-119092501437_1.html.

Grimnes, S., & Martinsen, O. G. (2000). *Bioimpedance and bioelectricity basics*. Cambridge: Academic Press.

Hell Stefan, W. Toward fluorescence nanoscopy. *Nature, 21*, 1347–1355.

Hench, L. L., & Jones, J. R. (2005). *Biomaterials, artificial organs and tissue engineering*. Cambridge: Woodhead Publishing Ltd.

Henry Royce Institute. https://www.royce.ac.uk/.

Hohman, M., Gregory, K., Chibale, K., Smith, P., Ekins, S., & Bunin, B. (2009). Novel web-based tools combining chemistry informatics, biology and social networks for drug discovery. *Drug Discovery Today, 14*, 261–270.

Hu et al. (2016). 3D-engineering of cellularized conduits for peripheral nerve regeneration. *Scientific Reports, 6*, 32184.

Hunsberger, J., Harrysson, O., Shirwaiker, R., Starly, B., Wysk, R., Cohen, P., Allickson, J., Yoo, J., & Atala, A. (2015). Manufacturing road map for tissue engineering and regenerative medicine technologies. *Stem Cells Translational Medicine, 4* 130–135. https://doi.org/10.5966/sctm.2014-0254.

iData Research. https://idataresearch.com/.

IIT Kharagpur To Run MBBS Course From 2021 Academic Year—NDTV. https://www.ndtv.com/education/iit-kharagpur-to-run-mbbs-course-from-2021-academic-year-2120345?amp=1&akamai-rum=off.

IMPacting Research, INnovation and Technology IMPRINT India Initiative. https://imprint-india.org/.

India Medical Devices Manufacturers, Distributors and Exporters Directory. (2011). http://www.pacificbridgemedical.com/publications/india/India_Medical_Devices_Directory_2011.php.

Innovation Knowledge Progress. https://www.ikpknowledgepark.com/ikp-group.html.

Inoue, H., Shin-eiKudo, & Shiokawa, A. (2005). Technology insight: Laser-scanning confocal microscopy and endocytoscopy for cellular observation of the gastrointestinal tract. *Nature, 2*, 31–37.

Istituto di Scienza e Tecnologia dei Materiali Ceramici. https://www.istec.cnr.it/en/.

Juette, M. F., Gould, T. J., Lessard, M. D., Mlodzianoski, M. J., Nagpure, B. S., Bennett, B. T., Hess, S. T., & Bewersdorf, J. (2008). Three-dimensional sub-100 nm resolution fluorescence microscopy of thick samples. *Nature Methods, 5*(6) 527–529.

Kalam Institute of Health Technology. https://kiht.in/.

Keeler, J. (2005). *Understanding NMR spectroscopy*, 2nd ed. Hoboken: Wiley.

Kohn, J. & Kantor, C., Devore, D. (2005). *Building a roadmap for the biomaterials science and technology to serve military needs*.

Kubitscheck, U. (2013). *Fluorescence microscopy: From principles to biological applications.* Wiley-Blackwell.

Langer, R., & Tirrell, D. Designing materials for biology and medicine. *Nature, 428*(6982), 487–492.

Liang, Peng, & Pardee, Arthur B. (1992). Differential display of eukaryotic messenger RNA by means of the polymerase chain reaction. *Science, 257,* 967.

Lichtman, J. W., & Jose-Angel-Conchello, J. (2005). Fluorescence microscopy. *Nature, 2,* 910–919.

Lippincott-Schwartz, J., & Manley, S. (2009). Putting super-resolution fluorescence microscopy to work. *Nature, 6,* 21–23.

Market Research Future. https://www.marketresearchfuture.com/.

Materials Assembly and Design Excellence in South Carolina (MADE in SC). https://nsf.gov/news/factsheets/southcarolina_factsheet.pdf.

McDowell, D. L., Panchal, J. H., Choi, H.-J., Seepersad, C. C., Allen, J. K., & Mistree, F. (2009). *Integrated design of multiscale, multifunctional materials and products.* New York: Elsevier.

McDowell, D. L., Panchal, J. H., Choi, H.-J., Seepersad, C. C., Allen, J. K., & Mistree, F. (2010). *Integrated design of multiscale, multifunctional materials and products.* New York: Elsevier.

McDowell, D. L. (2007). Simulation-assisted materials design for the concurrent design of materials and products. *JOM Journal of the Minerals, Metals and Materials Society, 59,* 21–25.

McGowan Institute for Regenerative Medicine. https://mirm-pitt.net/.

Medical devices are not drugs—Indian Express. https://indianexpress.com/article/opinion/columns/medical-devices-are-not-drugs-6086785/.

Medical Devices Regulatory Challenges in India. http://www.expresshealthcare.in/201112/strategy01.shtml.

Mitragotri, S., & Lahann, J. (2009). Physical approaches to biomaterial design. *Nature Materials, 8*(1), 15–23.

Muirhead, K. A., Horan, P. K., Poste, G. Flow cytometry: Present and future. *Nature, 3,* 337–356.

National Institute of Standards and Technology. https://www.nist.gov/.

NITI Aayog National Institute for Transforming India, National Health Stack: Strategy and Approach, 2018.

Niti Aayog proposes separate regulator for medical devices—The Economic Times. https://m.economictimes.com/industry/healthcare/biotech/healthcare/niti-aayog-proposes-separate-regulator-for-medical-devices/articleshow/71798635.cms.

Nolan, J. P., & Sklar, L. A. (1998). The emergence of flow cytometry for sensitive real-time measurements of molecular interactions. *Nature, 16,* 633–638.

Orive, G., Anitua, E., Pedraz, J. L., & Emerich, D. F. (2009). Biomaterials for promoting brain protection, repair and regeneration. *Nature Reviews Neuroscience, 10*(9), 682–692.

Panchal, J. H., Kalidindi, S. R., & McDowell, D. L. (2013). Key computational modelling issues in integrated computational materials engineering. *Computer Aided design, 45,* 4–25.

Pawley, J. (2006). *Handbook of biological confocal microscopy.* Berlin: Springer.

Pelczar, M. J., Jr., Chan, E. C. S., & Krieg, N. R. (2004). *Microbiology.* New Delhi: Tata McGraw-Hill Publishing Company Limited.

Perfetto, S. P., Chattopadhyay, P. K., & Roederer, Mario. (2004). Seventeen colour flow cytometry: Unravelling the immune system. *Nature, 4,* 648–655.

Place, E. S., Evans, N. D., & Stevens, M. M. (2009). Complexity in biomaterials for tissue engineering. *Nature Materials, 8*(6), 457–470.

Prometheus Group, KU Leuven. https://www.mtm.kuleuven.be/Prometheus.

Shorey, R., & Zarabi, M. J. (2011). *Technologies for healthcare sector in India.* Indian National Academy of Engineering, 2011, New Delhi, India. (https://www.inae.in/core/assets/fortuna-child/img/Healthcare.pdf)

Ramalingam, M. (2012). *Integrated biomaterials for biomedical technology.* Hoboken: Wiley.

Rust, M. J., Bates, M., & Zhuang, X. (2006). Sub-diffraction-limit imaging by stochastic optical reconstruction microscopy (STORM). *Nature, 3,* 793–795.

Sabareeswaran, A., Basu, B., Shenoy, S. J., Jaffer, Z., Saha N., & Stamboulis, A. (2013). Early osseointegration of a strontium containing glass ceramic in a rabbit model. *Biomaterials* 34, 9278–9286.

Saiki, R. K., Bugawan, T. L., Horn, G. T., Mullis, K. B., Henry, A. (1986). Erlich

Scheme for Promotion of Academic Collaboration. https://sparc.iitkgp.ac.in/index.php.

Schmittgen, T. D., & Livak, K. J. (2008). Analyzing real-time PCR data by the comparative CT method. *Nature, 3,* 1101–1108.

Sharma, V. K., Klingelhofer, G., & Nishida, T. (2013). *Mössbauer spectroscopy: Applications in chemistry, biology, and nanotechnology*. Hoboken: Wiley.

Siebert, P. D., & Larrick, J. W. (1992). Competitive PCR. *Nature, 359,* 557–558.

Silver, F. H., & Christiansen, D. L. (1999). *Biomaterials science and biocompatibility*. USA: Springer.

Smith, E., & Dent G. (2005). *Modern Raman spectroscopy: A practical approach.* Hoboken: Wiley.

Subach, F. V., Patterson, G. H., Manley, S., Gillette, J. M., Lippincott-Schwartz, J., & Verkhusha, V. V. Photoactivable mCherry for high-resolution two-color fluorescence microscopy. *Nature, 6,* 153–159.

Kalidindi, S. R., Medford, A. J., & McDowell, D. L. (2016). Vision for data and informatics in the future materials innovation ecosystem. *Journal of The minerals, Metals and Materials Society, 68,* 2126–2137.

Kalidindi, S. R., & De Graef, M. (2015). Materials data science: Current status and future outlook. *Annual Review of Materials Research, 45,* 171–193.

Technology Information, Forecasting and Assessment Council (TIFAC), Department of Science and Technology (DST), Technology Vision 2035: Technology Roadmap Materials, TIFAC, 2016, New Delhi.

The Materials Genome Initiative Strategic Plan. (2014). *Materials Genome Initiative National Science and Technology Council Committee on Technology Subcommittee on the Materials Genome Initiative.* https://www.whitehouse.gov/sites/default/files/microsites/ostp/NSTC/mgi_strategic_plan_-_dec_2014.pdf.

The Medical Device Regulation Bill. (2006). India. http://www.dst.gov.in/whats_new/whats_new07/MDRA-Act.pdf.

The Minerals, Metals & Materials Society (TMS), Building a Materials Data Infrastructure: Opening New Pathways to Discovery and Innovation in Science and Engineering, TMS, 2017, Pittsburgh

Vallet-Regi, M. (2014). *Bio-Ceramics with clinical applications*. Hoboken: Wiley.

Vermes, I., Haanen, C., Steffens-Nakken, H., & Reutelingsperger, C. (1995). A novel assay for apoptosis flow cytometric detection of phosphatidylserine expression on early apoptotic cells using fluorescein labeled Annexin V. *Journal of Immunological Methods, 184,* 39–51.

Vijayaraghavan, P., Mutte, H., & Sawant, H. (2016). *Indian Orthopedic Devices Market: A $2.4 billion opportunity*. Sathguru Management Consultants Pvt. Ltd. 2016

Waigh, T. A. (2007). *Applied biophysics—A molecular approach for physical scientists*. Hoboken: Wiley.

Wake Forest Institute for Regenerative Medicine. https://www.wakeforestinnovations.com/.

Wan, T. T. (2006). Healthcare informatics research: From data to evidence-based management. *Journal of Medical Systems, 30,* 3–7.

Warren, J. (2014). T*he materials genome initiative, data, open science, and NIST. 2014*. Research Data Alliance. https://rd-alliance.org/sites/default/files/RDA-MGI-ODIpptx.

Webster, T. J. (2012). *Nanomedicine technologies and applications*. Woodhead Publications.

White, J. G., & Amos, W. B. (1987). Confocal microscopy comes of age. *Nature, 328,* 183–184.

Williams, D. F. (1999). *The williams dictionary of biomaterials*. Liverpool, Great Britain: Liverpool University Press.

Williams, D. F. (1981). *Fundamental aspects of biocompatibility*. Boca Raton: CRC Press.

Williams, D. F. (1983). *Biocompatibility of clinical implant materials*. Boca Raton: CRC Press.

Williams, D. F., & Roaf, R. (1973). *Implants in surgery*. Philadelphia: W.B. Saunders.

Williams, D., & Zhang, X. (2019). *Definitions of biomaterials for the twenty-first century*. Elsevier. Materials Today.

Williams, D. F., Cahn, R. W., Haasen, P., & Kramer, E. J. (1992). *Materials science and technology: A comprehensive treatment medical and dental materials*. Wiley VCH.

Wyss Institute for Biologically Inspired Engineering. https://wyss.harvard.edu/.

Yuste, R. (2005). Fluorescence microscopy today. *Nature, 2*, 902–904.

Zhang, S. (2003). Fabrication of novel biomaterials through molecular self-assembly. *Nature Biotechnology, 21*(10), 1171.

Printed in the United States
by Baker & Taylor Publisher Services